AF616231

GENETIC ENGINEERING AND NEW POLLUTION CONTROL TECHNOLOGIES

GENETIC ENGINEERING AND NEW POLLUTION CONTROL TECHNOLOGIES

by

James B. Johnston and Susan G. Robinson

Institute for Environmental Studies
University of Illinois at Urbana-Champaign
Urbana, Illinois

NOYES PUBLICATIONS
Park Ridge, New Jersey, USA

Sole distribution by:
Gothard House Publications
Gothard House
Henley-on-Thames, Oxon
RG9 1AJ Tel. 049 12 3602

Library of Congress Catalog Card Number: 83-22077
ISBN: 0-8155-0973-1
ISSN: 0090-516X
Printed in the United States

Published in the United States of America by
Noyes Publications
Mill Road, Park Ridge, New Jersey 07656

10 9 8 7 6 5 4 3 2 1

Library of Congress Cataloging in Publication Data

Johnston, James B.
Genetic engineering and new pollution control technologies.

(Biotechnology review ; no. 3) (Pollution technology review ; no. 106)
Bibliography: p.
Includes index.
1. Sewage- -Purification- -Biological treatment.
2. Genetic engineering. I. Robinson, Susan G.
II. Series. III. Series: Pollution technology review ; no. 106.
TD755.J64 1984 628.3'51 83-22077
ISBN 0-8155-0973-1

Preface

This report concerns genetic engineering and waste treatment. It is the result of a planning study performed for the United States Environmental Protection Agency's Advanced Environmental Control Technology Research Center at the University of Illinois. The purpose of the study was to provide background information to a broad group of scientists, engineers and decision makers, so that opportunities for the improvement of biological waste treatment could be identified and evaluated. The report offers a core of information drawn from a very wide range of disciplines relevant to the genetic approach to new pollution controls. It identifies policy issues, research problems and scientific resources that bear on pollution abatement through genetic engineering.

In addition to assessing the potential for developing new pollution treatments, this report is intended to promote interdisciplinary communication among the specialized scientists and research managers engaged in such developments.

In so doing, the report intends to diminish one of the main impediments to the development of practical improvements in biological pollution treatment. Interdisciplinary communication is promoted by including background subsections describing basic disciplinary problems, techniques and other relevant information. The glossary and index should also be helpful.

It is believed that the information in this report will be particularly useful to professional engineers and research managers involved with pollution treatment problems. Molecular biologists may also find the background information in this report to be useful, especially the parts dealing with biological waste treatment, the natural limits to biodegradation and the establishment of microorganisms in the environment. Throughout this report, an effort has been made to provide references to relevant review literature, as well as to selected primary reports of use to specialists.

Because of the dynamic nature of the subject, this report cannot be considered to be comprehensive or complete. New questions continue to arise as pollution problems are recognized and as the technology of gene and microbe manipulation opens up new possibilities for addressing old problems. This report provides a reasonable outline of the state of the subject in mid-1983. It should be considered a guide for planning current work, hopefully to bring to practice the advanced pollution abatements that are the promise of the new genetic technologies.

NOTICE

Although the information described in this report has been funded wholly by the United States Environmental Protection Agency under assistance agreement EPA Cooperative agreement CR 806819 to the Advanced Environmental Control Technology Research Center, it has not been subjected to the Agency's required peer and administrative review and therefore does not necessarily reflect the views of the Agency and no official endorsement should be inferred. Mention of trade names or commercial products does not constitute endorsement or recommendation for use by the Agency, the authors, or the publisher.

Executive Summary

The objectives of this study were to document the basis for developing new pollution controls using genetic technology, to describe the present state of such developments and to recommend a research policy for such an approach. The study was performed by assembling three review papers dealing respectively with pollution problems, gene manipulations and natural limits to biodegradation. A panel of experts representing disciplines that might contribute to the development of new treatment technologies assembled in Urbana, IL to discuss these papers. Their recommendations were merged with the information gathered in the background papers, to provide a sound basis for recommending policy.

It was generally concluded that the spectacular recent advances in gene manipulation afford a remarkable new opportunity to design and create microorganisms with defined, desirable biodegradative capacities. But it was also concluded that this ability represents only one of the necessary conditions for the creation of new and practical pollution abatement processes. The other necessary condition is to bring a population of microorganisms having a suitable amount of a relevant catabolic potential into contact with a pollutant. This involves either establishing a new population of microorganisms in a polluted environment or distributing the genes for a biodegradative pathway among the members of an existing microflora at a polluted site.

Basic knowledge of these two processes is fragmentary compared to knowledge of gene manipulation technologies. Consequently the primary recommendation of the study is to support basic research on organism establishment and gene transmission in natural microbial populations. This recommendation states that the establishment of microbes, or their genes, in polluted matrices is as important to practical treatment processes as the creation of microbes with novel biodegradative capacities.

Release to the environment of "engineered" microorganisms was identified as an issue of concern to the public and to bodies such as the Recombinant DNA Advisory Committee (RAC), that is charged with the oversight of some of the new genetic technologies. To allay fears and to ensure a prudent and responsible development of new biological pollution treatments, it is recommended that experiments involving the release of genetically manipulated microorganisms be reviewed and approved for environmental safety by a scientifically qualified public body. When possible, organisms destined for release should be constructed without the use of *in vitro* recombinant DNA. This should expedite approval of release, since all techniques except the *in vitro* recombinant DNA method are currently viewed as essentially natural and innocuous.

The study identified a small number of research topics that are currently undersubscribed and that are likely to contribute substantially to new pollution treatments if research support were made available. These include research on the basic mechanisms underlying microbial co-metabolism and oligotrophy, the development of molecular genetics in filamentous fungi, in strict anaerobes and in archaebacteria, the study of the directed evolution of enzymes and metabolic pathways, and studies to advance understanding of dehalogenations by microbes. Two conventional treatment problems that might benefit from exploration from a genetic perspective were the removal of ammonia by low-aged wastewater sludges and improvements in the flocculation of these sludges. Unlike research on organisms establishment and gene transmission in indigenous populations, no priorities were assigned to these recommendations.

The study recognized that the advancement of waste treatments through genetic manipulations will require the involvement of scientists in a very broad range of scientific disciplines, ranging from the purely basic to the purely applied. Even in assembling the background papers for this study, communication among scientists was impeded by differences in disciplinary perspectives, in terminology and most importantly, because of a poor understanding of the fundamental capabilities and limits of the essential contributing disciplines. A number of measures were recommended to improve communication among the specialists in the contributing sciences. These include support for problem-focused symposia, an occasional newsletter, the coordination of research support among government agencies and a listing of pollution problems, research needs and current developments.

Table of Contents

1

Introduction

Purpose and Organization of the Study

This research planning study was conceived to assess the potential for developing new pollution control measures through advanced biotechnology. Its goals were to document the scientific basis for the genetic engineering approach to pollution control, to describe the current state of such developments and to recommend a research policy to promote it. The study began by assembling three background papers dealing with the following subjects: 1) a description of pollution problems potentially amenable to the genetic engineering approach, 2) a description of the genetic engineering technologies available for the creation of new solutions to these problems, and 3) a selective description of factors limiting natural biodegradative processes, some of which might be manipulated genetically to create new pollution control technologies. These papers were distributed to a panel of nationally recognized experts that later met at the University of Illinois to participate in a workshop organized around a discussion of these subjects. Each paper was also reviewed by a number of outside scientists and engineers who did not participate in the workshop (Table 1.) The discussion and reviews provided a basis for recommendations addressing the problems and opportunities that were identified. Three previous publications (Johnston and Robinson 1981; 1982a and b) described portions of this information and the views of the authors at various stages of the study's development. This publication constitutes the final report of the study.

The background papers were originally prepared as a means to span the broad range of information that might contribute to new pollution treatment technologies by providing a common core of knowledge intelligible to experts from all contributing disciplines. In this report the core information is provided in subsections labelled "Background: ...". These sections will present a non-specialist with sufficient information to understand the discussion in the text and, in addition, will give references to more comprehensive and specialized articles in the scientific literature. An attempt has been made to relate the background information to the practical aspects of pollution control. Nevertheless, specialists will probably find the background information dealing with their subject area to be elementary and they may wish to skip over such subsections.

The report is organized around the subjects of the original background papers. Sections 2-4 present and discuss this information in relation to pollution abatement technology. Section 5 discusses the release and fate of

Table 1
Reviewers of the Background Papers

Participants in the Urbana Workshop	
Martin Alexander	Cornell University
Peter Chapman	University of Minnesota
Stanley Dagley	University of Minnesota
David Gibson	University of Texas
I. C. Gunsalus	University of Illinois
Reino Kallio	University of Illinois
Morris Levin	U.S. EPA, Washington
Jonathan Raymond	CETUS Corporation
John Sykes	University of Sheffield
Kenneth Timmis	Max Planck Institut, Berlin
Albert Venosa	U.S. EPA Municipal Environmental Research Lab
David Watkins	U.S. EPA Industrial Environmental Research Lab
Outside Reviewers	
Charles Atwood	U.S. EPA, OHMS Branch
Edwin Barth	U.S. EPA Municipal Environmental Research Lab
Robert Bruener	CETUS Corporation
Leslie Glick	GENEX Corporation
Larry Gold	SYNERGEN Corporation
M.T.S. Hsia	University of Wisconsin
Homer M. LeBaron	CIGA-GEIGY Corporation
George Pierce	Battelle Columbus Laboratories
Jeanne S. Poindexter	Public Health Research Institute, New York
Gary Sayler	University of Tennessee
Robert Seely/James C. Linden	Biogas of Colorado
James Tiedje	Michigan Stage University
Moshe Uziel	James M. Montgomery, Inc.
Thomas G. Zitrides	POLYBAC Corporation

genetically manipulated microorganisms. Section 6 discusses the conclusions derived from the study, emphasizing the information gathered at the workshop and from the written reviews of the background papers. An attempt has been made to incorporate all of the comments of the reviewers into this report. The authors are responsible for any inadvertent omissions, as well as for the arrangement and relative emphasis given to each topic and opinion.

What is "Genetic Engineering"?

The purposeful alteration of the genetic information of an organism became so easy with the discovery of a means to recombine specific fragments of deoxyribonucleic acid (DNA) *in vitro* that the term "genetic engineering" began to be used to denote this activity. As with most new uses for well known words, the meaning of "genetic engineering" depends primarily upon the context created by the user. The term first referred to the rearrangement and interconnection of physically isolated DNA fragments, followed by the return of the newly arranged DNA to a living microbe. This process was also called "gene stitching". In current use the term "genetic engineering" has been extended by some to include the return of the rearranged DNA to higher life forms, such as green plants, where return has been accomplished in a limited number of cases, and mammals, where it has not yet been accomplished. Traditional methods of purposefully altering genetic information in higher organisms, by breeding or by mutation and selection, are usually not included in "genetic engineering" by these users. Microbiologists on the other hand will often include mutation, selection and all possible means of gene recombination in their meaning of "genetic engineering". All of these activities are commonly used to purposefully alter genetic information in microbes.

Confining the meaning of "genetic engineering" to "recombinant DNA technology" does not resolve all of the ambiguities created by current usage, since recently discovered means of mediating gene recombination *in vivo* with transposable elements and by such artificial means as protoplast fusion and protoplast transformation should also be considered forms of "recombinant DNA technology". Suggestions that terms such as "applied genetics" replace "genetic engineering" have merit in that they allow the inclusion of all of the traditional and new means of manipulating genes, but for the same reason this term loses the impact that many users wish to convey with "genetic engineering" because it refers to too broad a range of manipulations.

In this report, only microbial genes are discussed. Therefore, the term "genetic engineering" will include the purposeful modification of the genetic constitution of a microorganism by any means; it will be used synonymously with "applied genetics". Moreover, distinction will be made between *in vitro gene recombination* which involves the application of restriction endonucleases and DNA ligase to physically isolated DNA, and *in vivo gene recombination* which includes all rearrangements that occur in living organisms.

Pollution Control and the Genetic Engineering Industry

In the 1970s an expectation of revolutionary technological developments through genetic engineering was created by a sensationalist press. These expectations were abetted by a genuine series of spectacular advances in molecular biology and by a flood of private investment in new enterprises built around gene manipulation. The high expectations of the private sector were piqued by the immediate potential of the products of cloned genes in the pharmaceutical market. The supreme court's decision permitting the patenting of a living bacterium putatively useful for pollution abatement further stimulated private investment, since it appeared that patent protection might now be extended to genetically altered organisms. There followed an explosion of investment that was based mainly on unrealistic projections of the immediate commercialization of the new technology. At present, many of the new gene manipulation companies are retrenching, some have failed, but at the same time the first pharmaceutical products are coming to the market from a few of the new companies.

The application of genetic engineering methods to pollution problems has not been at the forefront of privately sponsored research. Many of the new biotechnology companies have decided that the near term return on investment from the development of new pollution treatment processes was not comparable to that anticipated by developing products for the pharmaceutical, chemical, agricultural or other markets. They have directed their efforts accordingly. Others among them were discouraged by the prospect of producing organisms for biological waste treatment commensurate with the sensational expectations reported in the press and competitive with similar products of biotechnology companies predating the discovery of the *in vitro* recombinant DNA method. For these reasons, the development of organisms potentially useful in new pollution treatments has been an activity supported primarily by the public sector.

2

Pollution Problems Potentially Amenable to Genetic Engineering

Introduction

To identify the pollution problems that might be attacked from a genetic perspective, scientists and managers were interviewed at the USEPA's Industrial Environmental Research Laboratory (IERL) in Cincinnati, the Municipal Environmental Research Laboratory (MERL) in Cincinnati, and the Hazardous Spills Unit in New Jersey; USEPA personnel in the Washington office were also contacted. Those interviewed were asked to name current and prospective pollution problems of widespread occurrence or having a high deleterious impact. These problems were then generically categorized. At this stage no assessment of the probability of achieving solutions through applied genetics was made. Assessment was left to the reviewers and workshop participants as experts in genetic technology. The problems identified are listed in Table 2, along with indications of a major example of each problem. This listing may be useful within the scientific community in identifying opportunities for the application of genetic techniques.

The problems identified in Table 2 are classified in three groups: aqueous waste streams, *in situ* treatment of polluted matrices, and miscellaneous. Two kinds of problems are found in the table, those involving biodegradation that are therefore perceived as amenable to genetic engineering, and some having a high deleterious environmental impact that are included because they currently have no economically feasible solutions, whether involving biodegradation or not.

The aqueous waste streams category refers first to potential improvements in the aggregate performance of conventional industrial and municipal wastewater treatment systems. The suggested improvements might involve decreases in the time needed for effective treatment, production of effluents with lower biochemical oxygen demand (BOD) and decreases in waste sludge production. Each of these measures of treatment performance represents the sum of thousands of reactions performed by a wide variety of microbes acting on a vast range of substrates. As will be discussed in the last section, improvements in these aggregate measures of treatment are unlikely to be achieved in the near future because the genetic manipulations currently possible affect only a few biochemical reactions at a time. Thus, they are unlikely to be attempted on the scale sufficient to affect overall performance.

Table 2
Problems Considered for Solution by Applied Genetics

1. Aqueous Waste Streams

   a. Improvements in aggregate measures of biological treatment, including decreased treatment time, improved effluent BOD, reduced sludge volume

   b. Specific improvements in treatment performance

      i. Acclimation of sludge for degradation or detoxication of a particular component of the waste stream, e.g., TCDD, haloaromatic compounds, organometallic compounds, etc.

      ii. Improved NH_3 removal, i.e., nitrification in low-aged sludge

      iii. Improved phosphate sequestration, e.g., using the luxury uptake of phosphorus

      iv. Improved flocculation and settling, i.e., reduced sludge bulking

2. _In situ_ Treatment of Pollution

   a. Contaminated soils, e.g., PCB's on North Carolina roadsides

   b. Contaminated sediments, e.g., Kepone in the James River

   c. Marine oil spills

   d. Hazardous waste landfills and their leachates, e.g., Love Canal

3. Miscellaneous

   a. Regeneration of physical absorbents, e.g., Kepone-laden activated carbon

   b. Specific diagnostic assays for components of wastes

   c. Biological scrubbers for air pollutants, e.g., SO_2

The acclimation of a complex microflora to a single target compound, on the other hand, may be responsive to currently feasible manipulations even though many kinds of organisms may need to be altered. In a wastewater treatment system, such acclimation might include either the restoration of effective treatment following an upset in the system or the effective destruction of a compound previously untreatable by the system. In both cases the underlying presumption is that a significant proportion of the organisms in the system become able to metabolize the target compound. For recalcitrant compounds, improvements in treatment imply a novel metabolism, presumably one developed by deliberate manipulation of organisms in the laboratory. The development of novel metabolisms is discussed in the section on gene manipulations.

To foster acclimation the organisms accomplishing the desired degradation must first be seeded into the system. They and/or their genes then must proliferate until a population of organisms is attained that can effectively catalyze the desired transformation. The proliferation of organisms may occur readily under some circumstances but may also be limited by the factors discussed in the section on the limits to biodegradation. Gene proliferation, on the other hand, might be accomplished in the absence of a concomitant proliferation of organisms if the relevant genes reside on a highly conjugative plasmid that can be transmitted rapidly throughout an indigenous population of microbes. This possibility was specifically mentioned as a plausible means whereby the seeding of a biological system could be rapidly effective. Similar proposals can be made for the acclimation of the microflora of a soil or sediment. Plasmid spread, not yet experimentally validated in sludges, soils or sediments, will be discussed in a later section.

There are many compounds that might be the target of attempts to improve biodegradation. In general, these include most slowly degraded and/or toxic substances. Compounds that are both recalcitrant and toxic deserve greatest attention. For instance, the penta- and tetrachlorodibenzodioxins (TCDDs) and their dibenzofuran analogues (TCDFs) would probably be ranked highest owing to their persistence (Matsumura and Benezet, 1982) and extreme toxicity (Cattabeni _et al._, 1978; Kimbrough, 1980; McKinney and McConnell, 1982). These might be followed by polychlorinated biphenyls (PCBs), the toxaphenes, Dieldrin/Aldrin, Heptachlor and its epoxide, the Chlordanes Mirex, Kepone and the hexachlorocyclohexane insecticides, among many others. In each of these cases microbial strains capable of rapid detoxication of these substances must be developed before acclimation of a biological system can be discussed. Many other toxic components of aqueous wastewater streams are less persistent than those and occur more frequently in typical industrial wastewaters. Phenols, anilines, nitro- and chlorobenzenes, nitro-, chloro-, and aminonaphthalenes and naphthols are examples. Bacterial strains are known that can detoxify or mineralize these compounds, and in some cases conjugative plasmids are known that encode appropriate degradative pathways (discussed in Slater and Bull, 1982). These compounds can therefore be used to evaluate attempts to acclimate a complex microflora, whether by proliferation of organisms or their plasmid-borne genes.

A number of pollution control companies currently offer microorganisms for sale as seed for the acclimation of operating treatment facilities. These commercial products are claimed to improve the treatment of specific components of wastewater streams such as phenols, greases or even "dioxin herbicide". If effective, nothing separates these products from the kinds of agents envisioned above. However, unambiguous experimental data demonstrating the effectiveness of these products in normally operating, full scale treatment systems having an active resident microflora has not been found in the refereed scientific literature. (One of these products has been found to establish an active microflora and perform effectively in a bench scale aeration basin lacking a resident flora [Holladay _et al._, 1978].) A body of published reports exists in the unrefereed literature that purport the effectiveness of other products. Most of the reports known to be available claim to show improvements in treatment in facilities seeded with the products. However, analysis of the data shows either that the improvements are unsubstantiated or that the experimental controls used were inadequate to clearly demonstrate whether the changes that took place were a result of the addition of the commercial seed, or were due to any of several uncontrolled factors. In addition, no clear description of the original derivation of the microbes is given in these reports. Consequently, it is not clear whether the organisms were developed through genetic manipulation or whether they were isolated from nature. It would be in the interest of these companies and the basic research establishment to know whether any of these products can accomplish the improvements claimed. From the information examined, it was impossible to determine whether any of them were beneficial, detrimental or irrelevant to treatment.

Improvements in ammonia removal, phosphate sequestration and flocculation were considered possible to the extent that these processes have a biological basis responsive to genetic intervention. The following background sections attempt to identify the biological factors involved in these processes and to point to a potential for genetically based improvements.

Background: Phosphorus Removal

Concern over the role of phosphate as a pollutant came during the late 1960's when public awareness was focused on the accelerated eutrophication of many natural water systems. Because of this concern, methods to remove phosphate from effluents were examined, the most common being the addition of lime to form a calcium phosphate precipitate. This method has the disadvantage of producing a voluminous chemical sludge that is difficult to de-water (Gaudy and Gaudy, 1980).

Observations of biological sludges at certain wastewater treatment plants revealed accumulations of substantial amounts of phosphates in the absence of chemical treatment. The phosphates were sequestered in the sludges formed in the aeration basins and were released when the same sludges were subjected to anaerobic conditions. These observations were the basis for the design of a treatment plant in Seneca Falls, NY, that removed phosphates from the effluent during aeration, followed by release into a small volume of water during settling (Fuhs and Chen, 1975). A chemical mechanism for this phosphate accumulation has been advanced, postulating that under aerobic conditions (high pH) the calcium, magnesium, and iron normally found in hard water will form insoluble phosphate precipitates. Under the conditions of low oxygen and pH that prevail during settling, the phosphate redissolved (Riding *et al.*, 1979).

Laboratory simulations of phosphate accumulation, however, point to the possibility of biological mechanisms of accumulation under some circumstances. For instance, 2,4-dinitrophenol, an uncoupler of oxidative phosphorylation, can reduce phosphorus uptake into biological sludges, suggesting a dependence on metabolism. Another theory with experimental backing suggests that phosphorus is removed from wastewater when the carbon to phosphorus ratio becomes so high that phosphate is the limiting nutrient. At least one report has been interpreted in this way. At the Jones Island Sewage Treatment Plant (Milwaukee, WI), brewery and domestic wastewater are treated together. A noticeable decline in phosphorus removal took place during a brewery strike, apparently due to the lack of the carbon-rich brewery waste in the influent, changing the C:P ratio in an unfavorable direction (Riding *et al.*, 1979). Normal domestic wastewaters, however, are usually too poor in carbon to effect phosphate removal as a growth-limiting nutrient.

Another biological mechanism for phosphate sequestration is called "phosphate over-plus" and its practical impact on phosphorus removal is controversial even among those holding the biological perspective. It involves a transient sequestration that occurs when phosphate-starved microorganisms are suddenly transferred to a phosphate-rich medium. While this phenomenon has been reproducibly demonstrated in the laboratory, its importance in real systems is usually discounted since no such shift occurs in typical wastewater treatment systems (Fuhs and Chen, 1975).

The final biological mechanism postulated for phosphate removal is called the "luxury uptake" of phosphorus (Gaudy and Gaudy, 1980). This mechanism suggests that certain strains of bacteria accumulate polyphosphate storage products (volutin granules) when starved for essential elements such as nitrogen or sulfur. Intracellular volutin granules have been observed in pure cultures of *Aerobacter aerogenes*, *Acinetobacter lwoffi*, and *Zooglea* species often found in activated sludges (Cosgrove, 1977). The *Acinetobacter*, isolated by Fuhs and Chen (1975) from sludge, was able to accumulate polyphosphate even in a complete medium containing adequate nitrogen and sulfur. This organism accounted for the majority of the observed phosphate uptake in two treatment plants. In a third plant at San Antonio, TX, highly efficient phosphate removal was found to be due in part to the presence of an unidentified gram negative coccus which accumulated volutin granules (Cosgrove, 1977). Low efficiency treatment plants seeded with this organism experienced a transient increase in phosphate removal efficiency.

Experimental demonstration of chemical and biological removal of phosphorus has been achieved. However, it is not yet possible to unequivocally ascribe phosphate removal in individual treatment plants to either purely chemical or biological mechanisms. In reality, each of these mechanisms may account for the behavior of phosphorus at any time under the changing conditions present in real treatment systems.

The isolation of new organisms able to sequester phosphate in the presence of adequate sulfur and nitrogen (Shoda *et al.*, 1979) gives hope that improvements in the efficiency of phosphate removal by microorganisms is possible. If such organisms, whether derived by genetic manipulations or not, can be established in wastewater treatment systems, then improvements in phosphate removal from domestic wastewaters may be achieved. Establishment of such strains may be aided by modifying the design of treatment plants to provide an environment conducive to the proliferation of the organism.

Background: Ammonia Oxidation

The oxidation of ammonia to nitrate (nitrification) is readily carried out by two groups of autotrophic bacteria that convert ammonia to nitrite and nitrite to nitrate, respectively. *Nitrosomonas* and *Nitrobacter* are genera typical of each of these groups (Gaudy and Gaudy, 1980). Oxidation of ammonia in wastewater is desirable since this can substantially reduce the nitrogenous oxygen demand often measured in the test BOD. In municipal wastewater treatment facilities that carry out nitrification, the process occurs predominantly in aeration basins having relatively high sludge ages, i.e., greater than 10 days. These long ages are necessary if the sludge is to contain a significant amount of the autotrophic nitrifiers, since these bacteria grow much more slowly than heterotrophs and cannot maintain themselves in a system when sludge is wasted at the high rates typical of secondary treatment steps. In practice, therefore, nitrification requires either an unusually large aeration basin or two separate basins, one of low sludge age in which organic carbon is oxidized by fast growing heterotrophic bacteria and one of high sludge age containing the nitrifiers.

From an economic perspective, it would be desirable to achieve nitrification in a more diverse and faster growing population of organisms, so that separate or unusually large basins would not be necessary, and so that a greater reliability could be achieved through the greater diversity of nitrifying organisms. Whether or not nitrification is obtained in a more diverse group of organisms, the minimum requirement for developing such advantageous organisms is that the nitrifying bacteria reproduce at rates comparable to heterotrophic bacteria so that they maintain themselves in the biological system.

Rapidly growing nitrifiers might be created by introducing the genes responsible for nitrification into heterotrophic bacteria normally resident in low age sludges and/or by accelerating the growth rate of autotrophic nitrifiers in some way. The division rate of an autotroph might be raised by relieving obligatory CO_2 fixation or by enabling them to derive useful energy from the oxidation of organic carbon available in the wastewater. Circumvention of the genetic regulation of cell division would also be necessary if this is limiting. Although many problems potentially mitigate against the success of any of these approaches to developing rapidly growing nitrifiers, it is not currently possible to assess the severity of these problems since necessary basic information relating to the genetics and metabolism of autotrophic nitrifiers is not available. The first practical step should be to obtain the basic information so that the efficacy of the measures mentioned above may be evaluated.

Background: Flocculation

The performance of a typical secondary wastewater treatment depends heavily on the efficiency of separation of the biomass from the treated water. For many municipal wastewaters, the clarification of secondary effluent limits overall performance (Palm *et al.*, 1980). Yet only extremely inexpensive separation methods are economically justifiable in most systems; typically the separation takes place in a sedimentation basin (clarifier) in which gravity settling of the microbial flora is followed by decantation or equivalent operation to remove the overlying liquid. If settling does not occur or occurs too slowly, the organisms can be washed out of the system, producing a turbid effluent and making the biomass unavailable for recycle to the influent wastewater. In extreme cases, the system can break down allowing untreated wastewater to pass through the system to pollute the receiving water.

Efficient settling of sludge is believed to be promoted by the interaction of two kinds of organisms, floc formers and filamentous bacteria (Sezgin, 1980). *Zooglea* and *Sphaerotilus* are typical genera performing these two functions. Floc formers are those that produce exudates and surface structures that promote the adhesion of microbial cells to each other forming clumps of hundreds to thousands of cells. Filamentous bacteria grow in a long fiber-like sheath that can cross-link the clumps produced by the floc formers. Efficient flocculation occurs when there is an optimum amount of floc formers and filaments in the sludge. When too few filaments are present, the flocs are not efficiently cross-linked and they tend to wash out, producing a turbid effluent (also called pin-point floc). An experimental parameter predicting the settling behavior of activated sludge is the total length of filaments present, determined microscopically and calculated per ml of sludge. Optimum settling occurs when less than 10^7 μm of filament occur per ml, in sludged with 700 to 4800 mg/ℓ suspended solids (Sezgin, 1980). When the filament length exceeds 10^7 μm per ml, the filaments themselves appear to support the structure of the flocs and prevent efficient compaction (bulking sludge).

The extended length of filaments is sensitive to conditions such as dissolved oxygen, organic loading and perhaps nutritional balance (iron, nitrogen, phosphorus) in addition to the more obvious characteristics such as reactor design (continuous or plug-flow), reactor operation (settling time, mixing, aerator type, etc.), and shear stress on the culture. Thus, the main advances in controlling flocculation have come from understanding and manipulating the microbial environment to favor a stable, desirable ratio and amount of floc formers and filamentous bacteria (Sezgin, 1980; Palm *et al.*, 1980).

Whether genetic manipulation of either of these kinds of microbes could improve flocculation under favorable conditions or maintain normal flocculation under adverse conditions is open for study. The genetic and biochemical factors regulating microbial adhesion and the growth of filamentous bacteria in response to their environment seem to be the two factors offering potential for genetic intervention. It is believed that no exploration of the molecular biology of filamentous bacteria has been undertaken. The adhesion of bacteria, on the other hand, has received the attention of biochemists and molecular biologists, although primarily within the context of the fouling of surfaces and of gene exchange by conjugation, a process involving cell-to-cell contact (Ellwood *et al.*, 1979; Berkeley *et al.*, 1980). There is opportunity to characterize all of these processes from the perspective of improving sludge flocculation. However, too little is currently known to realistically envision genetic manipulations that would yield improvements. Here, as with ammonia removal, a good deal of basic information needs to be obtained before modern molecular genetics can be brought to bear effectively.

In situ Treatments

In situ treatment is a much broader category of potential application than that described under aqueous wastewater streams. It involves applying an organism(s) to soils, sediments, or other media contaminated with persistent, toxic compounds. As with sludge acclimation, the organism or its genes must then proliferate so that a population develops that can effectively catalyze the destruction, immobilization or detoxication of the toxicant(s). Examples of polluted areas where this technique might be used include the miles of roadside in North Carolina contaminated by illegal dumping of PCBs, the Kepone-laden sediments of the James River, marine oil spills, spills of hazardous materials on land, or the area contaminated by toxic wastes in and around improperly constructed hazardous waste landfills. In discussing these problems with USEPA personnel, it was clear that they placed no restraints on the consideration of solutions to these problems. For most of them, there currently exist no economically feasible solutions and there may in addition be public pressure to find solutions.

In situ pollution problems are frequently so severe that proposed solutions need not necessarily be confined by the usually assumed attributes of biological treatments that make them inexpensive and convenient. These attributes include rapidity of action, growth of the microbial agents at the expense of compounds in the environment, and the mineralization of the toxicants by the microbe. A hypothetical treatment of a toxicant *in situ* might involve an organism constructed to partially transform and detoxify the noxious compound, but which will not grow and divide under the environmental conditions that prevail. For instance, this organism may produce very high amounts of a few enzymes responsible for transforming the target compound but which are of no actual benefit to the organism itself. Indeed, it may be necessary to fertilize the site of application of the organism with specific substances that it is designed to utilize, but which would not nourish the resident flora, thereby selectively supplying it with a source of energy to perform the desired transformation. Other chemical agents might also have to be applied to ensure the engineered strain's temporary persistence at the site in competition with an established resident flora.

Such an engineered strain differs from those mentioned in regard to the acclimation of complex microflora because the hypothetical organism in this case is actually a sophisticated biochemical catalyst rather than a living component of the microflora. However, since the hypothetical treatment does not involve the persistence of the engineered organisms nor is it as inexpensive as typical biological treatments, the problem addressed must have a compensatingly greater cost than typical problems, in order to justify the greater expense of producing, applying and feeding the organism at the site. A hypothetical organism that might at first glance justify such expense would be one catalyzing the dechlorination of penta- or tetrachlorodibenzodioxins or the analogous dibenzofurans, forming their tri- or lower chlorinated homologues. The acute toxicity of the homologues is very much less than that of the parent compounds (McConnell and Moore, 1978; Poland *et al.*, 1979; McKinney and McConnell, 1982). Such a treatment will also reduce the lipophilicity of these pollutants, potentially reducing their bioaccumulation and enhancing their biodegradation; so far the biodegradability of the lower chlorinated homologues of TCDD and TCDF have not been determined (di Domenico *et al.*, 1980; Philipp *et al.*, 1981).

The *in situ* problem mentioned most frequently by USEPA personnel was the detoxication of materials in and around toxic waste landfills. This problem was most often mentioned with the words, "is there anything that can be done?" The problem engenders strongly deleterious local and social impacts. Any new solutions should receive at least cursory consideration even if they are very novel and require considerable basic and applied research to achieve. Proposals involving biodegradation should acknowledge, however, that the physical constraints on microbes in these landfills can be very different from those in most other polluted environments. This stems from the variety and high concentration of the toxicants at these sites. Often the microenvironment in which the wastes occur is extreme, for instance having an unusually low water activity and relatively high concentrations of toxic halogenated compounds and heavy metals. In terms of their impacts on microbes, the microenvironments present in a toxic waste landfill can vary markedly and are usually less well characterized than those in ordinary soils or sediments.

Miscellaneous

Advanced biological treatment systems might also be made more useful if combined with physical agents. Some contaminated parts of the Kepone manufacturing plant at Hopewell, VA, were detoxified by a process using activated carbon, which itself became a contaminated medium. Disposal of this carbon by incineration at temperatures high enough to ensure destruction of the Kepone is technically feasible, but would be expensive and the carbon would be lost. If microbes could be developed that could attack the Kepone on the carbon, the carbon could be bioregenerated, potentially at great savings.

The use of carbon to trap pollutants, followed by biodegradation of the toxicants on the carbon, has been attempted with a variety of municipal and industrial effluents with encouraging results (Wilson, 1981; Suidan *et al.*, 1980). The bioregeneration of activated carbon is attractive for biological as well as economic reasons. Microbes able to attack recalcitrant or unusual organic compounds are expected to be specialists that will not tolerate rapid changes in environment, especially changes in the concentration of the target pollutants. Activated carbon adsorbs incoming organic compounds at the first available sites, tending therefore to be rich in adsorbed organics at the entrance to the column and poorer at the end of the bed, at least until all of the high affinity sites are occupied and breakthrough occurs. One method of carbon bioregeneration involves seeding a nearby saturated carbon bed with a mixed microbial flora and then recycling an aerated aqueous salt medium through it to supply oxygen and inorganic nutrients. Any organisms able to attach to the carbon and to grow at the expense of the adsorbed organic compounds are thereby provided with a relatively constant environment. If the carbon column was not saturated, the organisms have access to a wide range of ambient concentration of the target compound, increasing the chances that they will encounter a usable concentration. As organisms utilize the pollutant and deplete its local concentration, the organisms and their progeny are able to colonize new regions of the carbon, growing into areas with suitable concentrations of the pollutant.

The use of microorganisms in wet scrubbers for stack gases was speculated upon, for example, for the removal of SO_2 by conversion to sulfate. The speculation involves organisms that are both thermophilic and tolerant of the conditions present. One candidate might be the heat-and-acid-tolerant bacterium *Sulfolobus acidocaldarius* which has already been tested as an agent to desulfurize coal (Kargi and Robinson, 1982a and b). This species is capable of oxidizing sulfur to sulfate at pHs from 1 to 3 and temperatures from 60° to 85°C (Shivvers and Brock, 1973; Brock *et al.*, 1972). It is capable of heterotrophic growth in media such as yeast extract, but its capacity to oxidize sulfite is untested. Whether any plausible use of such a bacterium could be made in a wet scrubber depends upon many as yet unexplored facets of its metabolic capabilities, as well as on the physical and engineering constraints inherent in the contact between such an organism and combustion gases. These rather extreme speculations are mentioned here to draw attention to an unconventional approach to air pollution abatement, and to illustrate the very broad range of biological attributes that genetic engineering might manipulate to achieve desirable ends.

A few general suggestions were made for the use of genetic engineering methods on problems peripheral to pollution abatement. It was suggested that specific, rapid and convenient diagnostic tools might be developed using engineered organisms. Such strains might transform or utilize only a specific compound occurring in a wastewater and thus be suitable for auxanographic assay of it. These assays might be simpler or cheaper than chemical assays since separation of the compound might not be necessary. The auxanographic assay of vitamins and other cellular biochemicals in complex mixtures has been a routine method for decades. There are a great number of compounds found in wastewaters that might potentially be monitored in this way, including the priority pollutants. The question for the molecular biologist/biochemist is how to develop a strain that can utilize only a specific pollutant and do so reproducibly when this pollutant occurs as one component of a complex and potentially hostile mixture of compounds. A variation on auxanography might involve an engineered strain that transformed a specific target compound to a more easily quantified derivative.

Summary

All pollution problems solvable by enhanced biodegradation should be considered for evaluation when discussing new pollution treatments through genetic engineering. Several kinds of enhanced biodegradation are frequently desired. One is new or novel degradations. Here microbes are envisioned that can rapidly degrade a compound hitherto known only to be degraded very slowly, if biodegraded at all. Other desired enhancements are in the rate or extent of biodegradation of a target pollutant in an environmental setting. This enhancement requires that the organisms best suited to function in a particular environment become able to catalyze a desired biodegradation. In the most difficult problems, for instance PCBs in Hudson River sediments, both kinds of enhancements may be needed to bring about decontamination.

Without evaluating the potential for success, this section has identified various specific problems and categories of problems that might be approached through genetic engineering. Basic background information on phosphate removal, nitrification and flocculation was presented, along with some speculation on how the solutions of these problems might be approached. The mechanism of plasmid spread through an indigenous microflora was mentioned as a plausible means to acclimate a wastewater sludge, soil, or sediment to degrade a target compound. More highly speculative uses of genetically engineered microbes were raised with regard to scrubbing stack gases and developing new diagnostic assays for important pollutants.

3

Gene Manipulation Methods

Introduction

This section reviews and discusses microbial gene manipulation within the context of developing new pollution control technologies. The advances in molecular genetics over the past 10 years now permit, at least in principle, unrestrained rearrangement of the genetic information resident in microbes. This creates opportunities for the development of microorganisms to accomplish tasks that are deemed desirable by the manipulator, including manipulations to enhance biotransformations of environmental pollutants. The modifications that are readily envisioned as useful include: 1) the amplification of enzyme levels in an organism by selection of constitutive mutants, by linkage of genes to strong promoters, by increases in the number of gene copies or by any combination of these; 2) the rearrangement of regulatory DNA base sequences controlling the expression of specific genes in response to specific stimuli; 3) the introduction of genes for new enzymatic functions into organisms that do not normally possess them; and 4) the modification of individual genes to alter the characteristics of individual enzymes, including substrate specificities, kinetic constants (K_m and V_{max}), or characteristics such as the optimum pH.

The techniques used to accomplish these changes include: 1) the *in vitro* recombinant DNA technologies; 2) *in vivo* methods such as transposon mutagenesis and other transposon-mediated gene manipulations; 3) normal sexual genetic exchange by means of transduction, transformation or conjugation; 4) protoplast fusion; 5) generalized and site-specific mutagenesis; and 6) automated selection of desirable mutants. Each of these techniques is discussed in the following subsections.

Useful Organisms

The construction of new and useful microbial biodegradations will usually involve the introduction of new DNA into a bacterial cell. Knowledge of all available means to exchange genes must therefore be gathered at an early stage in strain development. Detailed information on gene exchange is already available for those families of bacteria that have been used to lay the foundations of molecular biology. These are primarily enteric bacteria and a few others with pathogenic properties. Most of the bacterial genera noted for the biodegradation of environmental pollutants, however, are not well characterized in terms of gene exchange.

To provide a starting point for discussion, criteria were chosen for organisms potentially useful in pollution control and the literature was surveyed for information on gene exchange in these genera (Table 3). Bacteria potentially useful in pollution treatment were defined as those 1) normally present in soils, sediments, or wastewater; or 2) tolerant of extreme or toxic environments; or 3) known to possess either a wide range of biodegradative capabilities or an individual ability that is known to be useful. Important types of biodegradative capabilities viewed as useful for pollution control included the capacity to metabolize long chain or branched alkanes, degradation of tri- or polycyclic-aromatic hydrocarbons, and the ability to perform dehalogenation reactions. Although listed by genus, the taxonomic position of some of the genera given in Table 3 is uncertain, notably Alcaligenes and Achromobacter. Also, several such as Nocardia and Mycobacteriaum may contain species that are somewhat diverse.

Keeping in mind that any organisms used for pollution abatement will be released to the environment, additional criteria for useful organisms include non-pathogenicity to man, animals, or plants. An examination of Table 3 reveals several genera containing plant or animal pathogens. Since any genus may contain pathogenic and non-pathogenic species, rigorous characterization of any species under consideration for use in pollution control should be required to determine the extent of its relation to pathogenic members of the same genus.

Table 3 should be a good starting point for interrelating information on genetic exchange and properties useful for strain development. It is, however, impossible for such information to be complete. Not only will new information on gene exchange continue to appear, but the consideration of any particular pollution problem will suggest useful genetic information not included in the categories defined here. When devising the construction of a new strain, any source of appropriate genetic information can be used.

Gene Exchange and Biodegradation by Fungi

The fungi play an important role in the recycling of carbon in the environment. In acidic and cold habitats, such as the northern conifer forests of North America, they are more important to the recycling of carbon than the bacteria. In environments where bacteria are dominant, fungi are frequently present and they may play essential roles. Filamentous fungi are frequently a constituent of biofilms attached to solid surfaces, for instance in biofilm reactors in wastewater treatment facilities, and in the beds of eutrophic streams. Fungi living in association with tree roots in the rhizosphere play important roles in metabolizing plant exudates and in mobilizing inorganic nutrients. Fungi are well known decomposers of polymeric substances such as cellulose and lignin.

Analysis of genetically controlled functions in fungi goes back to the beginnings of modern genetics but detailed study at the molecular level has just begun, notably in yeasts such as Saccharomyces (Bacila et al., 1978). The metabolic potential of the fungi relevant to pollution control often occurs in filamentous species that are poorly explored from the genetic

Table 3: Genera Potentially Useful for Environmental Biotransformations

Genus	Known Means of Genetic Exchange	Reference	Degradative Activity	Reference
Achromobacter	transformation	Juni & Heym, 1980	dehalogenation	Bollen, 1960; Furukawa *et al*, 1979; Reutergardh, 1980; Ou & Sikka, 1977
	transduction	Twarog & Blouse, 1968	hydrocarbons	Jobson *et al*, 1972; Mulkins-Phillips & Stewart, 1974
Acinetobacter	lysogenic phage	Herman & Juni, 1974	PCBs	Furukawa *et al*, 1979; Furukawa *et al*, 1978
	transformation	Sawula & Crawford, 1972; Juni & Janik, 1969; Juni, 1972; Juni, 1978	hydrocarbons	Atlas, 1979
	conjugation	Towner, 1978; Juni, 1978	diethyleneglycol	Pearce & Heydeman, 1979
	transduction	Twarog & Blouse, 1968; Juni, 1978		
Agrobacterium	plasmids	Hooykaas *et al*, 1980; Hernalsteens *et al*, 1980	dehalogenation	Bollen, 1960; Jensen, 1957
	transposon	Hooykaas *et al*, 1980; Hernalsteens *et al*, 1980		
Alcaligenes	plasmid	Fisher *et al*, 1978	PCBs	Furukawa *et al*, 1979
			linear alkylbenzene sulfonates	Johanides & Hrsak, 1976; Bird & Cain, 1974
			hydrocarbons	Atlas, 1979; Kiyohara *et al*, 1982
			polycyclic aromatics	Atlas, 1979
			dehalogenation	Engesser *et al*, 1980
Arthrobacter	lysogenic phage	Casiday & Liu, 1974	hydrocarbons	Atlas, 1979
			polycyclic aromatics	Stevenson, 1967
			pentachlorophenol	Stanlake & Finn, 1982
Bacillus	plasmid	Stewart & Levin, 1977; Biamucci & Lovett, 1977; Tanaka & Koshikawa, 1977	long chain alkanes	Kachholz & Rehm, 1980
	transformation	Dubnau, 1976	aromatics	van der Linden & Thijsse, 1965
	lysogenic phage	Hutchison & Halvorson, 1980; Graham *et al*, 1979		
	protoplast fusion	Gabor & Hotchkiss, 1979; Fodor & Alfoldi, 1976		
	transduction	Bramucci & Lovett, 1977		

(continued)

Table 3: (continued)

Genus	Known Means of Genetic Exchange	Reference	Degradative Activity	Reference
Brevibacterium	(no known genetic exchange, genus of uncertain affiliation in Coryneform group)		dehalogenation aromatics polycyclic aromatics	Horvath, 1971; Horvath & Alexander, 1970a Atlas, 1979 Atlas, 1979
Clostridium	plasmid	Li *et al,* 1980	dehalogenation	Schuphan & Ballschmiter, 1972
Corynebacterium			dehalogenation	Bollen, 1960
Flavobacterium	(no known genetic exchange; genus of uncertain affiliation)		hydrocarbons	Jobson *et al,* 1972; Atlas, 1979
Hydrogenomonas	(genus reclassified as *Pseudomonas)*		dehalogenation	Pfaender & Alexander, 1972
Methanogenic consortium	(consortium members not characterized)		aromatics (anaerobic)	Evans, 1977
Micrococcus	(no known genetic exchange; genus in same family as *Staphylococcus)*		branched hydrocarbons hydrocarbons	Jobson *et al,* 1972 Carlberg, 1980
Moraxella	transformation	Juni, 1974; Juni, 1977	aromatics (anaerobic)	Evans, 1977
Mycobacteria	transduction	Ramakrishnan & Shaila, 1979; Gelbert & Juhasz, 1973; Gelbart & Juhasz, 1970; Twarog & Blouse, 1968; Capet-Antonini & Mankiewicz, 1968; Sundar Raj & Ramakrishnan, 1970	aromatics branched hydrocarbons cycloparaffins	Ratledge & Winder, 1966; Rogoff, 1961 van der Linden & Thijsse, 1965 Beam & Perry, 1974; Carlberg, 1980
	transfection	Tokunaga & Nakamura, 1968		
	pseudolysogeny	Baess, 1971		
	plasmids	Crawford & Bates, 1979		
	conjugation	Konicek & Konickova-Radochova, 1975		
	"mating" recombination	Tokunaga *et al,* 1973		
Nocardia	lysogenic phage	Crockett & Brownell, 1972	polycyclic aromatics	Carlberg, 1980
	"mating" recombination	Brownell & Adams, 1968; Schupp *et al,* 1975	hydrocarbons	Mulkins-Phillips & Stewart, 1974

(continued)

Table 3: (continued)

Genus	Known Means of Genetic Exchange	Reference	Degradative Activity	Reference
Pseudomonas	plasmids	Bagdasarian & Timmis, 1982; Bagdasarian *et al*, 1981; Sagai *et al*, 1977; Palchaudhuri, 1977; Hedges & Jacoby, 1980; Finger & Krishnapillai, 1980; Clark *et al*, 1977; Sagoo & Cain, 1978; Hopper & Kemp, 1980; Chakrabarty, 1976; Yano & Nishi, 1980; Gautier & Bonewald, 1980; Jacoby & Shapiro, 1977; Stanisich & Richmond, 1975; Franklin *et al*, 1981	dehalogenation	Slater *et al*, 1979; Slater & Bull, 1982
Pseudomonas	conjugation	Holloway *et al*, 1971; Holloway & Richmond, 1973; Stanisich & Richmond, 1975	polycyclic aromatics	Dunn & Gunsalus, 1973; Barnsley, 1975
	transduction	Chakrabarty *et al*, 1968; Stanisich & Richmond, 1975	hydrocarbons	Chakrabarty, 1976
	lysogenic phage	Stanisich & Richmond, 1975; Miller *et al*, 1974	heavy metals	Summers & Silver, 1978
	transposons	Jacoby & Shapiro, 1977; Chakrabarty *et al*, 1978		
Streptomyces	lysogenic phage	Okanishi & Okami, 1966	dehalogenation	Chacko *et al*, 1966
	transposons	Sermonti *et al*, 1980	diazinon	Gunner & Zuckerman, 1968
	protoplast transfection	Suarez & Chater, 1980; Okanishi *et al*, 1966a and b		
	transformation (plasmids)	Bibb *et al*, 1978		
	protoplast transformation	Krugel *et al*, 1980		
	protoplast fusion	Baltz, 1978; Hopwood & Wright, 1979		
Xanthomonas	conjugation	Kordyum *et al*, 1980	hydrocarbons	Atlas, 1979
			polycyclic aromatics	Atlas, 1979

perspective, for instance in Cunninghamella (Cerniglia and Gibson, 1977, 1979) and Cladosporium (Walker and Cooney, 1973). In the unicellular fungi most amenable to genetic analysis, much remains to be learned about the enzymology of degradations relevant to pollution control. An example of work exploring this area is the recent study of aromatic acid catabolism by the yeast Trichosporon (Anderson and Dagley, 1980).

The mechanics of genetic analysis in the filamentous fungi is more difficult than in unicellular forms, but the filamentous fungi should be a rich source of genes for catabolic enzymes. The genetic analysis of filamentous fungi at the molecular level is complicated by the variety of sexual and asexual reproduction exhibited by various species, the syncitial and multinucleate structures of many species, and by the paucity of proven methods to introduce specific segments of DNA into fungal cells. At present the filamentous fungi are considered primarily as a source of DNA that may encode novel or useful catabolic enzymes. It appears, however, that basic research on both pollution-related biotransformations and the basic molecular genetics of all of the fungi may make available the many other potentially useful attributes of these organisms, including tolerance of hostile environments, the ability to attack polymers and associations within the rhizosphere.

In the remainder of this report, the focus will be placed on bacteria. Presently the scope of possible gene manipulation with fungi other than yeasts is so very small that the brief description just given is adequate. Moreover, it is a conclusion of this report that an expansion of the gene manipulations possible with typical environmental fungi should be undertaken. With the yeasts, where extensive gene manipulation is already possible, it would be desirable to further explore the pollution-related biodegradative capabilities that they express.

Gene Exchange and Biodegradation by Archaebacteria and Anaerobic Eubacteria

Like the fungi, two broad categories of bacteria deserve to be developed as objects of genetic manipulations. The archaebacteria, considered by some to be a unique form of life (Fox et al., 1980; Magrum et al., 1978), are commonly found in anaerobic environments where they catalyze methanogenesis, the conversion of CO_2 and H_2 to methane. In these environments, they interact with the other category of promising bacteria, the anaerobic eubacteria, that break down organic matter and supply the methanogenic archaebacteria with their substrates (Wolfe, 1979).

These two classes of organisms dominate the biological activity in anaerobic habitats on earth. Such habitats are very important economically and in terms of the global carbon cycle. Anaerobic habitats are widespread and occur in marine and fresh water sediments, bogs, tundra, ruminant animals, and wastewater treatment facilities.

Purely from the perspective of wastewater treatment, anaerobic decomposition of organic matter is already widely practiced. The anaerobic digestion of activated sludge is a widely used means to reduce sludge volume, and certain kinds of anaerobic filters are also found in more advanced wastewater treatment plants. Yet, the biochemistry and genetics underlying these processes are essentially unknown. The most detailed biochemistry of anaerobic metabolism currently available comes from studies of methanogenesis by pure cultures (Wolfe, 1979). In the laboratory, mixed populations of anaerobes have shown the capacity to dissimilate benzoic acid (Ferry and Wolfe, 1976), catechol (Suidan et al., 1980), phenol (Suidan et al., 1981), and certain halogenated aromatic compounds (Suflita et al, 1982). The biochemical steps involved in these decompositions have not been elucidated. But these studies, as well as the favorable experience with anaerobic wastewater treatment, suggest a genuine potential for improvement through genetic manipulation.

The genetics of methanogenic bacteria and many of the anaerobic eubacteria are essentially unexplored. Attempts to isolate auxotrophic mutants of methanogens have met with difficulty, not least due to the apparent impermeability of their isoprenoid cell wall complex to exogenous growth factors (J. Konisky, unpublished experiments). Gene libraries of methanogen DNA have been constructed and the first attempts to identify specific gene products are in progress (Konisky; Reeve; unpublished experiments). But at present, there is no indication of normal gene exchange among the methanogens.

The other organisms frequently found in methane-producing environments are responsible for converting polymeric and other organic material to fatty acids, formate, acetate, CO_2, and H_2 (Wolfe, 1979). This group of organisms is poorly characterized from the microbiological, biochemical and genetic perspective. Clostridium and Bacteroides are two genera known to occur in this group, but many members remain unidentified. These organisms are believed to be responsible for the reductive decomposition of the aromatic and chlorinated compounds mentioned previously. It would be very desirable to understand the biochemical pathways that they employ, and to identify the genetic elements controlling these activities, in order to assess the potential for developing an attack on environmental pollutants.

Archaebacteria are also found in highly acidic and highly saline environments. Their unique cell envelope probably plays an important role in their tolerance of these extreme environments. Whether these archaebacteria possess, or could be made to acquire, biodegradative capacities useful in polluted and extreme environments is unknown. The intriguing possibility of introducing new DNA into such species is suggested by the recent report of plasmid DNA in Sulfolobus acidocaldarius, an acid- and heat-tolerant archaebacterium (Yeats et al., 1982).

Background: Plasmids, Especially Metabolic Plasmids

Plasmids are autonomously replicating, double stranded, circular, extrachromosomal DNA. Plasmids range in size from a few to a few hundred million daltons (Clowes, 1972; Hansen and Olsen, 1978; Bukhari et al., 1977). They can encode a very wide variety of genetic information. Those encoding the ability to exchange genetic information by conjugation (sex factors) were noticed first. Other biochemical functions encoded in plasmid DNA include drug resistances, toxin production and catabolic pathways. In nature the biological role of plasmids seems to be to provide a vehicle for the exchange of genetic information among bacteria.

The only essential attributes of a plasmid are circularity of the DNA and the presence of a site for the initiation of DNA replication. These two properties are often exploited by molecular biologists wishing to manipulate genes. By introducing a gene of interest into a plasmid, as described in later sections, the plasmid's replication origin is exploited for making copies of the gene. Circularity is exploited for the physical isolation of plasmid DNA including any genes that it may encode.

By submitting DNA extracts containing plasmids to density gradient centrifugation in the presence of an intercalating dye, covalently closed circular DNA, i.e., plasmids, can be separated from any DNA having free 3' and 5' ends, i.e., chromosomal DNA. Equilibrium density gradient centrifugation separates DNAs on the basis of density. The topology of the DNA, i.e., double-stranded circular or free-ended, alters the extent of binding of intercalating dyes. This results because intercalating dyes unwind the DNA double helix when they bind. Free-ended DNA can unwind to accommodate the dye but closed circular DNA resists unwinding. Intercalating dyes and DNA differ in density; thus DNAs that bind different amounts of dye, because of differing topology, give rise to dye-DNA complexes of differing density. The dye-DNA complexes made from free-ended and closed circular DNA separate according to their net density when subjected to equilibrium density gradient centrifugation.

Many kinds of genetic information can be associated with plasmid DNA. The ability to transmit bacterial genes by cell-to-cell contact (conjugation) when found is always associated with plasmids. Conjugation may involve the transmission of chromosomal genes as well as genes encoded on plasmids. Not all plasmids carry the genes for conjugation, but genes present on non-conjugative plasmids can be mobilized by conjugative plasmids coexisting in the same cell. Other biochemical functions associated with plasmid genes may render a plasmid's host bacterium resistant to toxic agents including antibiotics, heavy metals, UV radiation and toxins called bacteriocins. Resistance to bacteriocins is frequently associated with bacteriocin synthesis by the same cell. The biochemical functions that produce antibiotic and the other resistances are often encoded in one or a few genes in contrast to the dozens of genes needed for conjugation.

Plasmid genes can undergo reciprocal recombination and transposition _in vivo_ and therefore can readily and randomly exchange genetic information with their host's chromosome (see later sections). The natural importance of plasmid DNA appears to be the ability to distribute genetic information among a wide variety of bacteria by the combined effect of recombination, transposition and conjugation. Thus, plasmids permit the formation of a large number of combinations of genes including combinations that may be favored in a particular habitat. In this way they serve the evolutionary process by helping to create genetic diversity.

In contrast to properties conferred by one or a few genes, certain plasmids are known that encode extensive catabolic pathways, in some cases involving dozens of genes in several organized groups. These so-called metabolic plasmids have been implicated in the biodegradation of camphor (Rheinwald *et al.*, 1973), naphthalene (Dunn and Gunsalus, 1973; Zuniga *et al.*, 1981; Skryabin *et al.*, 1980; Dunn *et al.*, 1980; Connors and Barnsley, 1982), toluene-xylene (Williams and Murray, 1974; Friello *et al.*, 1976), unbranched alkanes (Chakrabarty *et al.*, 1973), lignin (Salkinoja-Salonen and Sundman, 1979; Salkinoja-Salonen *et al.*, 1979), *p*-chlorobiphenyl (Kamp and Chakrabarty, 1979), the herbicide chloridazon (Kreis *et al.*, 1981), halogenated phenoxyacetic acids (Pemberton and Fisher, 1977; Fisher *et al.*, 1978; Don and Pemberton, 1981), and halogenated alkanoic acids (Kawasaki *et al.*, 1981a, b). The existence of extensive and well organized plasmid pathways seems to be an exception to the view of plasmids as vehicles for the random pickup and exchange of single genes because it is hard to imagine the organizing of the DNA for such extensive pathways on a plasmid by a series of random gene rearrangements and acquisitions. This reasoning suggests that degradative pathways have evolved on the chromosomes of strains living in stable habitats, where the opportunity for such complex gene organization would be greater. Such strains might be the source of the pathways that have been found on plasmids. This expectation appears to be borne out by examples of plasmid-less strains catalyzing some of the same degradations that have been associated with certain metabolic plasmids (Connors and Barnsley, 1982) and by degradative pathways behaving as transposable elements, i.e., sets of genes able to move freely among replicating DNA's (Chakrabarty *et al.*, 1978).

Metabolic plasmids are interesting for pollution control both because they offer the potential to spread new genetic information among indigenous populations of microorganisms, leading to the expression of new biodegradations, and for the potential of manipulating plasmid genes in the laboratory to extend the range of substrates that the plasmid hosts may attack. Relevant manipulations include the piecing together of a new biochemical pathway by assembling a set of genes on a plasmid, and the creation of an altered enzyme activity by the site-directed mutagenesis of a plasmid gene.

Specific Means to Manipulate Genes: in vitro and in vivo Recombination

The newest means for manipulating the microbial genome involves *in vitro* and *in vivo* recombination. Recombination is the breakage and rejoining of DNA fragments to form a new base sequence. *In vitro* recombination refers to the breakage and rejoining of isolated DNA in the test tube, usually using purified enzymes such as type II restriction endonucleases and T4-DNA ligase. *In vivo* recombination pertains to DNA rearrangements that take place exclusively in living cells. Specific uses of these two techniques can accomplish remarkable feats of strain construction and genetic analysis, especially when coupled to modern methods of mutagenesis and strain selection.

The technique of in vitro recombination is recognized as a major addition to the array of available gene manipulation methods. It allows virtually unlimited rearrangement of DNA base sequences and in conjunction with methods to reintroduce chemically isolated DNA into living organisms, it permits unprecedented movement of genes within and among organisms. A few impediments to its use come from a lack of proven means to introduce isolated DNA into some species (Table 3) and barriers to the expression of introduced, foreign DNA in some host organisms. Both of these limitations are the subject of intensive study that has considerably narrowed their scope in the few years since in vitro recombination was first practiced. It can be anticipated that in time there should be no insurmountable barriers to the universal application of the in vitro techniques, at least in bacteria.

Background: in vitro Recombination

So much has been written about the technique of in vitro recombination that repetition of details of the method would be redundant. Excellent reviews are available (Murray, 1980; Vosberg, 1977; Low and Porter, 1978, Dart et al., 1981) as are papers in the primary literature describing important variations in technique, for instance in the use of DNA packaged in vitro into phage (Hohn and Murray, 1977; Loenen and Brammar, 1980), the use of synthetic recognition sites (Scheller et al., 1977), and blunt-end ligation (Scheller et al., 1977; Sgaramella, 1972; Wartell and Reznikoff, 1980). Improvements in technique continue to appear; in addition to the primary scientific literature, useful sources of information on the latest advances include the staff of departments of molecular biology at major universities and private companies manufacturing the enzymes, chemicals and apparatus used in these methods.

For readers completely new to in vitro recombination, the essence of the process is contained in the following steps and in Figure 1:

(1) Isolation of the Source DNAs. DNA isolation can be performed by a wide variety of methods varying in efficiency and in the degree of intactness of the molecules finally isolated. Often the genes of interest are linked to a plasmid or phage genome so that they can be replicated and recovered by the isolation of these DNAs. Protocols for the isolation of various kinds of DNA from microbial sources are found in reviews (Clowes, 1972), compendia of methods (Davis et al., 1980; Crosa and Falkow, 1981), and in individual specialized articles (Hansen and Olsen, 1978; Johnston and Gunsalus, 1977; Birnboim and Doly, 1979; Wheatcroft and Williams, 1981). It is generally assumed that DNA can be isolated from any living organism. Special difficulties may, however, be encountered when attempting to isolate a particular gene of interest present in a vast mixture of unwanted genes. Also, certain forms of DNA from some sources, for instance certain plasmids from Pseudomonas and Streptomyces strains, may be unusually difficult to isolate intact.

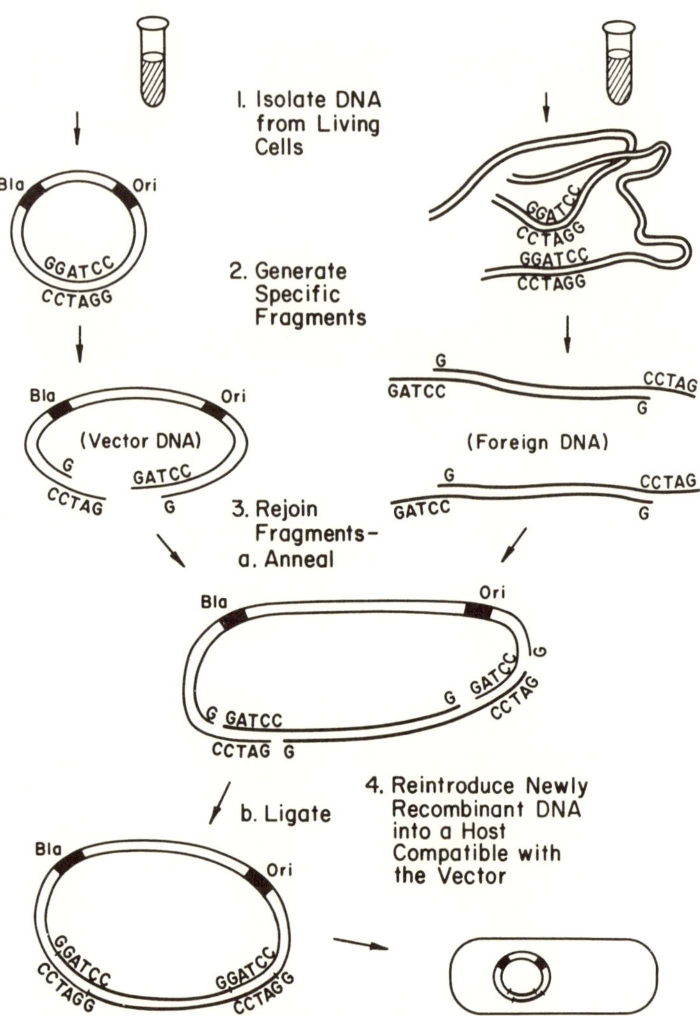

Figure 1. Schematic of the steps involved in *in vitro* recombination. The vector DNA in the figure is a plasmid bearing a gene for resistance to β-lactam antibiotics (Bla) and a replication origin (Ori). The vector also has a unique site for restriction endonuclease Bam H1.

(2) Generation of Specific DNA Fragments. This is one of the two essential steps of the method. Type II restriction endonucleases acting on isolated DNA are most often used to accomplish this step. Endonucleases are enzymes that hydrolyze the phosphodiester backbone of DNA within the molecule rather than at its ends. (Exonucleases degrade DNA at its ends.) Restriction endonucleases are those involved in self-non-self DNA recognition systems in bacteria and perhaps other organisms. They recognize a DNA sequence and hydroyze both strands of the DNA at specific sites. Type I restriction endonucleases cleave the DNA at sites remote from the recognized sequence while Type II enzymes cleave it within the recognized sequence (Nathans and Smith, 1975). Thus, any DNA fragment produced by a particular Type II enzyme will include part of the recognized base sequence at its ends whereas Type I fragments will not. In addition, any sequence recognized by a Type II enzyme is a palindrome, a sequence reading the same from front to back or back to front. For example, the enzyme BamHl recognizes the six-base sequence:

```
          ↓
5' — GGATCC — 3'
3' — CCTAGG — 5'
              ↑
```

and introduces cuts in each DNA strand where indicated by the arrows. This enzyme produces fragments with complimentary single-stranded ends:

```
— G           GATCC —
— CCTAG  +        G —
```

Not all restriction enzymes produce single-stranded ends; some cut leaving "blunt" ends, i.e., fragments ending in a complimentary base pair. Since Type II restriction endonucleases recognize specific palindromes, they will produce a specific set of fragments from any DNA that they digest. These fragments will directly reflect the number and location of the recognized sequence. DNA fragments can be separated according to size by electrophoresis in agarose or polyacrylamide-agarose gels. By this means individual fragments and the genes they encode can be isolated (Nathans and Smith, 1975).

(3) Rejoining of Specific DNA Fragments. This is the second essential step of the process. It is accomplished by mixing DNA fragments and resynthesizing the phosphodiester bond using a DNA ligase. If the fragments to be joined possess complimentary single-stranded ends, they might first be incubated at a suitable temperature and in a medium of suitable pH and ionic strength to favor association of the complimentary single-stranded ends (reannealing) before ligation. Blunt-ended fragments without complimentary ends can also be rejoined. However, this can only be accomplished using T4 DNA ligase (Sgaramella, 1972).

In some situations, it is desirable to join fragments obtained using different endonucleases, and hence having nonidentical single-stranded ends. The single-stranded ends generated by some pairs of restriction endonucleases are sufficiently homologous to permit reannealing and ligation even though the complimentarity of the ends is not perfect. It is also possible to blunt the ends of fragments and to join them by blunt-ended ligation. Blunting can be accomplished either by cutting off the single-stranded portion with S1 nuclease (a single-strand specific exonuclease) or by filling in the strand opposite the single strand using DNA polymerase (Ullrich *et al.*, 1977). Other methods to promote the joining of unlike-ended fragments may involve the addition of synthetic linker DNA fragments, such as polydA:dT single-stranded tails (Jackson *et al.*, 1972; Lobban and Kaiser, 1973) or synthetic linker sequences (Scheller *et al.*, 1977). Machines for the automated synthesis of gene fragments (Alvarado-Urbina *et al.*, 1981) facilitate this last method.

(4) Reintroduction of the Recombinant DNA (rDNA) into a Living Cell.

Formally, *in vitro* recombination is complete with step 3 above, but the great power of the method lies in the ability to reintroduce the rDNA into a living cell in such a way that it can be replicated and expressed. The means to introduce DNA into living cells include direct uptake by the cells (transformation and transfection), assembly of the DNA into a phage capsid *in vitro* with subsequent infection of cells by this phage (Hohn and Murray, 1977; Loenen and Brammar, 1980), uptake of DNA by protoplasts (protoplast transformation)(Hopwood, 1981; Chater *et al.*, 1982), including DNA previously contained in lipid vesicles (liposomes) (Dimitriadis, 1979; Hopwood, 1981), or related stimulatory preparations (Rodicio and Chater, 1982). Of these, only transformation is a wholly natural process, occurring typically in species such as *Bacillus*, *Streptococcus*, *Agrobacterium*, *Rhizobium* and *Acinetobacter* (Table 3). Transformation by circular or linear DNA can be artificially induced in many other species by treatment with calcium, heat shock or osmotic shock (Cohen *et al.*, 1972; Bergmans *et al.*, 1980).

Once introduced the rDNA must be replicated or it will be lost during cell division. Replication occurs if the DNA is incorporated into the chromosome of the cell by *in vivo* recombination or if the introduced DNA has a sequence for the initiation of replication suitable to the DNA replicases produced in the cell. Thus, when using the *in vitro* recombination methods, a plasmid or part of a phage bearing a DNA replication origin is frequently one of the pieces joined to the foreign DNA. These DNAs furnish the base sequences necessary for stable replication of the foreign DNA, including in some cases the gene(s) for the DNA replicases as well. Each bacterium that takes up and replicates a piece of foreign DNA can give rise to descendants that also bear the foreign DNA base sequence. These descendants are said to be clones of the original cell. The DNA fragment responsible for the autonomous replication of the new DNA in the host is referred to as the vector for the foreign DNA. Each such vector is usually used only with certain compatible host organisms.

Various means have been devised to recognize clones of host organisms that have taken up a vector, and to discriminate among these to find the clones whose vectors also bear a piece of foreign DNA. One widely used method employs plasmid vectors encoding two separate antibiotic resistance genes (Figure 2). By growing the transformed bacteria in the presence of the first antibiotic, only cells that have taken up the vector survive. Among these some will contain vector DNA that has incorporated rDNA while others will have only incorporated the original vector DNA. Usually the second antibiotic resistance gene on the vector harbors the restriction endonuclease site used in the *in vitro* recombination process. Therefore, any incorporated foreign DNA will be found inserted in this gene and the gene will be inactivated by this insertion. If it is intact, this antibiotic resistance gene will be functional and no foreign DNA has been inserted. By searching among clones resistant to the first antiobiotic (selection for uptake of vector DNA) for clones that have become sensitive to the second antibiotic due to inactivation of the resistance gene, the clones bearing the foreign DNA are found. Recently two new plasmid vectors have been described that permit a more direct selection of clones bearing inserted foreign DNA (Roberts *et al.*, 1980; Dean, 1981). There also exists an elegant phage vector/host system for identifying clones bearing foreign DNA. In this case the insertion of DNA in a particular gene allows the phage to replicate in a certain host not supporting phage replication if this gene is intact (Loenen and Brammar, 1980).

In vivo Recombination

Protoplast fusions and transposon-mediated gene manipulations are the two principal methods relying on *in vivo* gene recombinations. In each case, it is necessary for the manipulator to find the desirable gene combinations from among those that arise by random recombinations. As described in the following background section, transposons are especially useful in this process because clones bearing them are easily selected by means of the antibiotic resistance gene(s) that they usually carry.

Strains manipulated exclusively by *in vivo* recombination are subject to fewer regulatory restrictions than are strains manipulated by the *in vitro* methods and hence are advantageous for practical applications. The difference in regulatory scrutiny stems from the perception that *in vivo* gene rearrangements occur naturally and therefore create no unusual risks (Department of Health and Human Services, 1978).

Original Vector DNA:

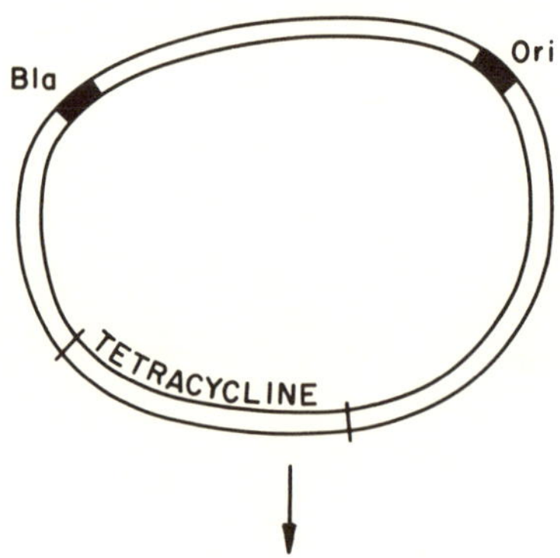

Recombinant DNA with Inactivated Tetracycline Resistance Gene

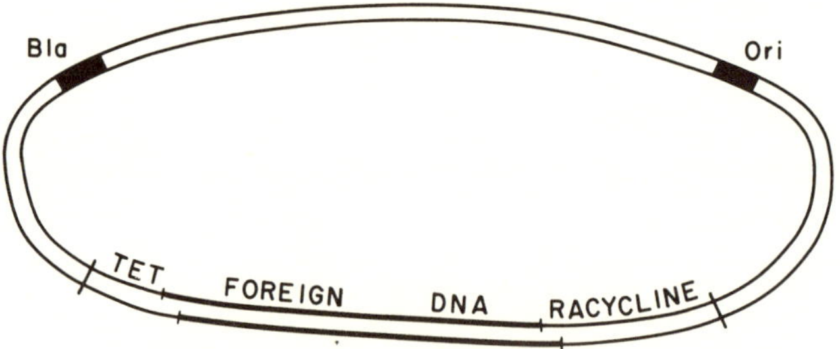

Figure 2. Inactivation of a gene by insertion of foreign DNA. In this illustration the sense of the word "tetracycline" is destroyed by the insertion of the words "foreign DNA". The functionality of the products in a gene (proteins and RNA) can be destroyed in a similar way when foreign DNA is inserted into a gene.

Protoplast Fusion

Protoplasts are viable bacterial cells that have lost their rigid peptidoglycan layer retaining only their cytoplasmic membrane. They are usually formed by treatment of dividing bacteria with glycine and lysozyme in an osmotically protective medium. Protoplasts fuse readily in media containing appropriate concentrations of polyethylene glycol when they are pelleted gently in a centrifuge (Baltz, 1978; Fodor and Alfoldi, 1979; Hopwood and Wright, 1978; Schaeffer, *et al.*, 1976; Hopwood, 1981; Chater *et al.*, 1982). After fusion genetic rearrangements can occur to give rise to new gene combinations. Any gene arrangement able to regenerate a vegetative cell can be recovered from the fusion and the desirable recombinants can be selected. The procedure is presently practiced primarily with Gram positive bacteria because of difficulties in regenerating Gram negative protoplasts. Protoplasts are also useful for achieving high efficiencies of gene transfer by transformation (Bibb *et al.*, 1978) and transfection (Suarez and Chater, 1980; Krugel *et al.*, 1980; Okanishi *et al.*, 1966, 1968) either with or without prepackaging of the DNA into liposomes (Hopwood, 1981; Dimitriadis, 1979; Rodicio and Chater, 1982).

Background: Transposon-mediated Gene Manipulations

Transposons are short DNA base sequences (2-20 kilobase pairs) able to insert *in vivo* into many sites in replicating DNA. Transposons are bounded at their ends by special base sequences, some of which are the "insertion sequences" originally found in the *gal* and *lac* operons and in certain phage (Shapiro *et al.*, 1977). Transposons that have been analyzed in detail have revealed a gene for an enzyme called transposase that is essential for transposition (Chou *et al.*, 1979; Gill *et al.*, 1979; Kostriken *et al.*, 1981). In addition, transposons frequently encode one or more nonessential genes that confer a selectable phenotype such as antibiotic resistance (Bukhari *et al.*, 1977), lactose degradation (Cornelis *et al.*, 1978), or aromatic hydrocarbon catabolism (Chakrabarty *et al.*, 1978).

Transposons have several properties that make them very useful for gene manipulation and genetic analysis (Figure 3) (Kleckner *et al.*, 1977). They can insert into a large number of sites in replicating DNA and can therefore be used to study almost any specific DNA region. Genes into which the transposon inserts usually suffer a total loss of function due to the disruption of the gene's base sequence (insertional inactivation, transposon mutagenesis). The function of a gene of interest can therefore be eliminated by dividing the gene's base sequence into two parts but without actually eliminating the base sequence of the gene itself. Transposon mutations behave as point mutations in transmission mapping, and consequently are very useful in two and three point crosses. Transposon mutations are strongly polar when they occur within an operon, i.e., the transcription of genes within the operon and beyond the site of transposon insertion is reduced.

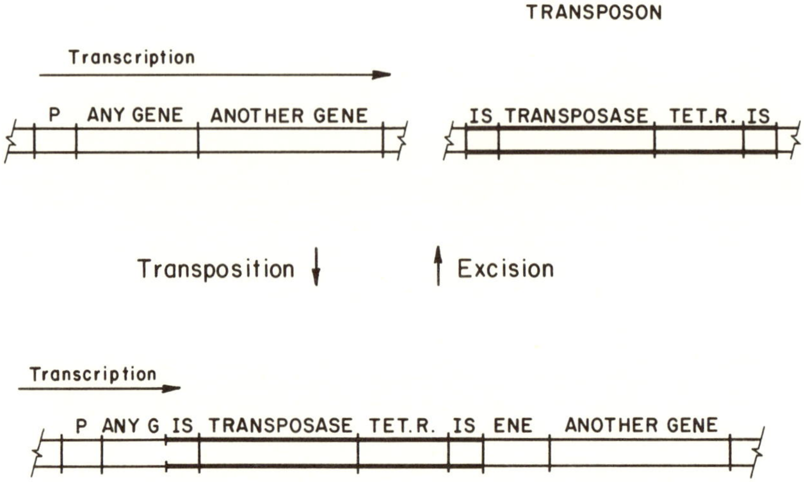

TRANSPOSON

Transcription

P ANY GENE ANOTHER GENE

IS TRANSPOSASE TET.R. IS

Transposition ↓ ↑ Excision

Transcription

P ANY G IS TRANSPOSASE TET.R. IS ENE ANOTHER GENE

Figure 3. Transposon insertion inactivates a gene and blocks transcriptions of other genes (polar mutations). In the original state, transcription of the operon on the left procedes from the promoter site (P) through two genes. After the transposon inserts into the first gene, that gene is destroyed and subsequent transcription of the other genes in the transcriptional unit is blocked. The transposon is bounded by insertion sequence (IS) and contains genes for a transposase and resistance to tetracycline.

Transposon-generated mutations are relatively stable but are capable of a precise, spontaneous reversion restoring the original base sequence of the mutated gene. These revertants are characterized by a loss of the transposon. Loss is not associated with the insertion of the transposon into a new site, but is a separate event.

The antibiotic resistance determinants carried by certain transposons provide a directly selectable phenotype for DNA regions into which the transposon has inserted. Thus, when a gene undergoes transposon mutagenesis, two simultaneous phenotypic changes are recognized, the loss of the function encoded by the mutated gene and the gain of the function, e.g., antibiotic resistance, encoded by the transposon. This makes the selection of desired mutants extremely easy. First, a population of cells is exposed to the transposon and plated on a medium selecting for transposon insertion, for instance by selecting antibiotic resistance. Then these cells are examined to find those that have lost the function of interest. Since typical bacteria only possess several thousand individual genes and since transposons insert randomly and rarely into a genome, mutant cells having lost almost any desired function can be found by directly assaying only a few thousand bacterial colonies for the missing function.

Another desirable consequence of transposon mutagenesis is that only a single gene in a cell is usually affected. Typical chemical mutagenesis often causes mutations in many other parts of the genome than the gene of interest. Such "silent" mutations can adversely affect the mutant obtained by chemical mutagenesis or strains constructed from it. Every transposon-derived mutant, in contrast, can be recognized by the transposon's selectable phenotype. Multiple transpositions in a cell can occur, but they are not frequent, and some transposons apparently inhibit second transpositions into the same replicon (Robinson _et al._, 1977).

Transposon Mutagenesis

The practical application of transposons for strain construction or genetic analysis first requires a means of introducing the transposon into the cell. If DNA bearing a transposon is introduced into a cell, the transposon sequence can then introduce itself randomly at a low frequency into any replicating DNA present. The first objective of an investigator is, therefore, to devise a means to create a population of cells in which the only stable transposons are those that have newly inserted into one of of the hosts' replicons. Several means have been used to accomplish this. Transposons present on phage DNA have been introduced by infection of host cells that will not support the replication of the phage (Kleckner _et al._, 1977). Only cells in which transposition has occurred will stably maintain the transposon in subsequent generations. The presence of the transposon is recognized by its selectable phenotype, usually an antibiotic resistance. In another method, a plasmid encoding a transposon is first introduced into a cell, allowing transposition onto the host chromosome or another plasmid resident in the cell. Wide host range plasmids such as RSF 1010::Tn1 (Heffron _et al._, 1975) are particularly useful for this purpose. The problem is then to separate the transposon-donor, e.g., RSF 1010::Tn1, from the rest of the genome so that cells in which transposition has occurred can be recognized by the transposon's selectable phenotype. This might be accomplished by a conjugation in which the transposon-mutagenized cell is the male parent and the female parent is a related strain carrying a transposon-less plasmid of the same incompatability group as the transposon-donor plasmid. The presence of this incompatible plasmid in the female parent reduces the frequency at which the transposon-donor plasmid will occur in transconjugants. Therefore selection of cells exhibiting the transposon's phenotype, for instance by growth in the presence of a specific antibiotic, will enrich for recombinant strains in which the transposon has inserted into a DNA different from the transposon-donor plasmid. An alternative scheme employs transposon-bearing plasmids unstable under certain environmental conditions. Plasmids temperature-sensitive for replication, e.g., pMB5 (Robinson et al., 1980) or pTH10 (Harayama et al., 1980; 1981), have been used to enrich for transposition by selecting for the transponson's phenotype while growing cells at a temperature that does not permit the replication of the original transposon-bearing plasmids. Certain Inc P1 plasmids bearing a transposon and a Mu phage have been found to be unstable in some hosts, notably Rhizobium (Beringer et al., 1978; Johnston et al., 1978). Conjugation of such plasmids from a host in which it is stably maintained to one in which it is unstable, while selecting for the transposon, permits the isolation of transpositions into the recipient's replicons.

All of these schemes describe the movement of a transposon from a vector to a host replicon. There exists in addition a general technique to isolate transposon insertions into conjugative plasmids. This requires transposition in the opposite direction, i.e., from the host chromosome to plasmid. Conjugative plasmids introduced into cells carrying transposons in the chromosome can be mated with transposon-less recipients and the exconjugant can be grown on a medium which selects for transposon-containing cells. Under these circumstances most exconjugants will carry plasmids bearing transposon insertions.

Other Uses of Transposons

Kleckner et al. (1977) have described a comprehensive array of transposon-mediated gene manipulations and this has been used as the core of a Cold Spring Harbor course on advanced bacterial genetics. Beyond the use of transposon-mutants in mapping and as a means of creating selectable mutations in or adjacent to a gene of interest, they describe the use of transposons for:

1. complementation tests
2. cloning and physical identification of genes
3. generation of Hfr strains with a specific origin and direction of transfer
4. construction of new F' episomes
5. construction of specialized transducing phage
6. production of deletions and duplications with predetermined endpoints

7. facile creation of a collection of deletion mutants useful for deletion mapping
8. site-directed chemical mutagenesis

The physical identification of genes is possible by mapping the location of transposon insertions in them. The physical mapping of transposon insertions is accomplished by comparing agarose gel electrophoretic patterns of the restriction endonuclease digests of the transposon-mutagenized and the unmutagenized parental DNAs. The molecular weight of the mutagenized fragment will increase by an amount corresponding to the transposon DNA. The transposon can then be more precisely mapped by electron micrography of a heteroduplex formed from the corresponding parental and mutagenized DNA fragments. Since the mutant was formed by insertion of the transposon into the gene of interest, the location of the transposon identifies a point within the base sequence of the gene.

One means of site-directed chemical mutagenesis utilizes a transposon to locate a region of DNA that has been exposed to chemical mutagen. Originally used in bacterial strains amenable to transduction, the method requires first that a transposon mutant be obtained with the transposon inserted in a gene adjacent to the gene to be chemically mutagenized. A transducing phage lysate is prepared on this host. Such a lysate will contain phage encapsulating host DNA fragments. These phage are exposed to a chemical mutagen, followed by infection of a normal host that is then grown on a medium selecting for the transposon. Only phage incorporating the transposon will give rise to transductants under these conditions. The transductants so isolated will frequently bear mutagenized DNA from the region adjacent to the site of the original transposon insertion, thereby providing a population of cells rich in mutations in the desired gene.

All of the transposon-mediated gene manipulations presume that one or more physiological means of gene exchange are available to introduce transposon-bearing DNA into the strain to be manipulated. Some manipulations such as the site-specific mutagenesis just described, also presume a knowledge of gene linkage. For many of the genera named in Table 3, such knowledge may only be available in small amounts. Therefore, transposons may be of limited value with these species until a more extensive genetic characterization has been completed.

It should be easy to supply the needed knowledge for eubacteria since genetic exchange by normal physiological means appears to be rampant among them. However the exchange of DNA among archaebacteria is essentially unexplored, and the unusual structure of the cell envelope in these organisms may portend equally unusual features concerning the exchange of DNA. The utility of transposon-mediated gene manipulations can be expected to grow as new characterizations of gene exchange in the eubacteria appear and as the archaebacteria are studied for the first time.

Gene Identification

In vitro gene manipulations often require identification of the physical segment of DNA encoding the gene to be manipulated. This has been accomplished by a variety of techniques, including transposon mutagenesis and mapping, just described. Each method has special requirements that impose corresponding restrictions.

The shotgun cloning experiment is a general technique often facilitating gene identifications. DNA fragments from a source including the relevant gene are randomly recombined with a recombinant DNA vector and subsequently transferred to a host organism. The gene of interest is then identified by screening clones of such transformed hosts. If the gene is expressed, the desired clone can be selected by the complementation of a mutation in the recipient strain. Clones of bacteria or phage plaques bearing the desired gene can also be identified by hybridization and autoradiography using a ^{32}p-labelled radioactive probe of complementary oligonucleotide (Grunstein and Hogness, 1975). This method moves gene identification back to the problem of obtaining a specific probe. Often probe oligonucleotides are isolated from specific m-RNA's by synthesis of a complementary DNA (c-DNA) using reverse transcriptase. (Reverse transciption can also be used to directly produce DNA for cloning, Little *et al.*, 1978). The sensitivity of the method using hybridization and autoradiography is high enough that fragments of DNA present in a single copy per cell can often be detected *in situ* in single bacterial colonies in typical plate cultures. Even so, the sensitivity of the method can be increased by cloning the foreign DNA into a vector whose concentration per cell can be increased. These so-called amplifiable vectors include plasmids similar to the colicin El-plasmid and lamda phage (Nakano and Masuda, 1982). Amplification of these plasmid vectors is accomplished by incubation of the cells bearing the recombinant plasmid with chloramphenicol, and for recombinant phage by multiple rounds of infection. Amplification provides much larger than normal amounts of the recombinant DNA per unit volume. Thus the ^{32}p-labelled probe DNA has more complementary DNA to bind to, thereby increasing the assay's sensitivity. Another method to identify clones bearing the desired gene(s) employs labelled antibody reacting with the product of a recombinant gene in *in vitro* extracts or in clones of host bacteria (Skalka and Shapiro, 1976; Kemp and Cowman, 1981).

Techniques using both *in vivo* and *in vitro* recombination methods have been used to indirectly identify some specific gene products. If a bacterium is mutated by insertion of a transposon into a gene of interest, or alternatively into a gene known to map very closely to a gene of interest, the DNA from such bacteria can be shotgun cloned and the genes subsequently isolated by selecting for clones expressing the transposon's phenotype (Kleckner *et al.*, 1977).

Gene Expression

A problem that may impede the functionality of some strains constructed using DNA from diverse origins is a lack of gene expression. Although many foreign genes are expressed in frequently used hosts, a number of exceptions are known. Genes from *Streptomyces fradiae* failed to complement a number of *E. coli* auxotrophic mutations (Horinouchi *et al.*, 1980); *E. coli* DNA bearing genes for several antibiotic resistance and auxotrophic functions was not expressed in *B. subtilis* (Ehrlick and Goze, 1979; Ehrlich and Sgaramella, 1978), and certain metabolic plasmid genes from *Pseudomonas putida* were not expressed in *P. aeruginosa* and *E. coli* (Chakrabarty *et al.*, 1978). In one particular case, a block in transcription of the foreign DNA was suggested as the point of inhibition (Chakrabarty, 1974).

Theoretically, the failure of any of the steps from transcription of DNA onward can pose a barrier to proper gene expression. Ehrlich and Sgaramella (1978) discussed the possibility for failures of transcription, RNA processing, translation and the processing of preproteins. He has also mentioned the possibility that the product of an rDNA gene may weaken or kill its host cell, for instance, by producing toxic intracellular metabolites not normally found in the host, or by decreasing the tolerance of the host to environmental toxicants, for instance, by enhancing permeation.

The problem of transcriptional barriers mentioned previously has been overcome, at least in certain hosts. Plasmid vectors have been constructed in which the rDNA can be inserted adjacent to a promoter-operator region that functions in the host, ensuring that the gene will be transcribed (Roberts *et al.*, 1979; Hallewell and Emtage, 1980). Plasmids of this sort have been derived from both the *lac* and *trp* operons of *E. coli*. In addition, one of these cloning vectors can be used to interpose any desired number of base pairs between the *trp* operator and the rDNA, a factor that modulates gene expression. Techniques similar to these can also be used to replace a weakly transcribed promoter with a strongly transcribed one, enhancing the production of the structural gene product. These manipulations may have application to enhance the amount of catabolic enzymes in an engineered strain or to place the expression of a constructed pathway under the control of man-made, exogenous compounds that are inducers of these operons.

Site-specific Mutagenesis

Knowledge of the detailed enzymology of DNA replication, recombination and repair has opened up methods for *in vitro* mutagenesis of isolated DNA at precise locations in the polynucleotide. Recent reviews (Shortle *et al.*, 1981; Smith and Gillam, 1981a,b; Timmis, 1981) provide excellent descriptions of the principles of these methods and references to the primary literature. Without repeating the information in the reviews, it can be pointed out that all of these methods require a means of returning the isolated DNA to a compatible host organism if the functionality of the mutation is to be evaluated. In some of the methods a rather detailed knowledge of the DNA base sequence at the site to be altered is also needed. Since such basic knowledge is frequently lacking some of these methods are currently of little value in bacterial species possessing catabolic pathways of high potential utility for pollution control.

Some site-specific mutagenesis methods only require the existence of a single-strand nick in an isolated DNA bearing the gene to be mutated. Such methods are broadly useful. They are more likely than other methods to contribute to the development of strains for pollution control since this prerequisite is easily met.

Once a mutated segment of DNA is incorporated and expressed, it is essential that a means be available to identify the desirable mutations of a gene among those generated. Within the context of developing strains for pollution abatement, the desirable mutations might enable a strain to metabolize a pollutant untouched by the parental strain, or might allow the strain to function better than its parent in a stressful environment. The great advantage of the site-specific method in this context is that only the DNA encoding one or a few select target genes is altered. In conventional mutagenesis many genes in a strain are altered simultaneously increasing the probability that silent or deleterious mutations may occur in a strain bearing a selected mutation.

Automatic Screening of Mutants

The automated selection of bacterial colonies having mutant phenotype has progressed significantly in recent years due in particular to the availability of microprocessor technology. Perhaps the epitome of automated selection technology is the cell sorting and identification machine designed by Glaser and his collaborator under the auspices of the National Institutes of Health (Sevastopoulos *et al.*, 1977). This device is capable of screening up to one million individual bacterial colonies on the basis of colony morphology, and has been used to detect and isolate a number of new temperature-sensitive cell division mutants of *E. coli*.

The key to automated mutant selections is the ability to recognize a desired clone by morphological or other machine-identifiable traits. A number of commercial and experimental plating and screening devices that handle a smaller number of colonies than the Glaser machine have been constructed. Typical among them are those described by Sharpe *et al.* (1978). Most of these devices are intended for the microbial segments of the pharmaceutical, medical and food industries.

As yet the automated screening of genetically manipulated cultures to isolate desirable variants has not been widely practiced. Also, most of the currently available devices screen colonies of organisms, not individual cells. Recently, flow cytometers with cell sorting capability have become available. These machines can detect and segregate individual particles with unusual dimensions. They have found use with cultured eukaryotic cells but have been applied to bacteria only to a limited extent (Steen and Boye, 1980).

Older technologies like continuous culture have been adapted for the "automatic" selection of mutants with desirable catabolic activities. Senior, Slater, Bull and their collaborators have applied this approach to isolate mutants catalyzing dehalogenation (Senior *et al.*, 1976;

Slater *et al.*, 1979). They grew a carbon-limited culture of *Pseudomonas putida* on propionate in the presence of a five-fold excess of 2,2-dichloropropionate, the herbicide Dalapon. Since the strains could not use the Dalapon, the propionate, called the carrying substrate, sustained the culture. Any mutants able to dehalogenate Dalapon and use it as a carbon source had a greater carbon pool available to it than those using the propionate alone and therefore displaced the parental type from the chemostat. Use of this isolation technique in concert with site-specific mutagenesis holds great promise for the adaptation of existing catabolic pathways to recalcitrant analogues.

Directed Evolution of Enzymes

The gene manipulation technologies described up until now have concentrated on methods for rearranging preexisting genes isolated from diverse organisms. They do not permit the creation of new enzymes catalyzing unique new reactions. A number of exploratory studies, however, have pointed to the possibility of deliberately altering the properties of enzymes through the directed evolution of specific genes under strong selective pressures. These methods may give new meaning to Reiner's declaration, "Some enzymes are more discriminating than others, but it seems fair to say that any enzyme can be fooled if one goes to enough trouble" (Reiner, 1959).

The first attempt to alter the properties of an enzyme in a deliberate fashion was carried out by Clarke (1974). She used various techniques to select mutant forms of a single growth-rate limiting enzyme in *Pseudomonas aeruginosa*. The mutants she sought would utilize various analogues of the enzyme's original substrate, acetamide. The amidase under study was a single polypeptide inducible in the wild type strain by acetamide and able to catalyze the hydrolysis of acetamide or formamide to the corresponding acid and ammonia. Clarke used culture conditions in which either the acid or ammonia was growth-limiting, and sought mutants using substrates related to the original. The original amidase could barely hydrolyze butyramide but could not attack higher acyl analogues of acetamide nor could it utilize any aryl amides. The study commenced with the isolation of spontaneous mutants able to grow at the expense of butyramide. By sequential rounds of selection for spontaneous or induced mutants and finally by using more than one amide as substrate, Clarke was able to isolate mutants that extended the specificity of the amidase enzyme to amides never used by the parental strain. Similar studies extending the metabolism of five-carbon sugars and alcohols in *Klebsiella* have been described (Hartley *et al.*, 1977; LeBlanc and Mortlock, 1971a and b; Oliver and Mortlock, 1971a and b; Mortlock, 1982).

In all of these pioneering studies a few common features emerge. The initial mutations were found in regulatory genes leading to constitutive production of the wild type form of the growth-rate-limiting enzyme. It happened in all cases that the enzyme could utilize the first substrate analogue being tested, but only at a lower rate than that of the normal substrate. In the amidase study the overproduction of the enzyme compensated for the inherently slow utilization rate and permitted a sufficient carbon flux to pass into cellular metabolism to support growth.

In two studies, subsequent mutations were found in the gene for the rate-limiting enzyme. These mutants specified an enzyme form with altered substrate specificity or altered K_m. Following several such sequential mutations, amidase mutants were finally obtained that readily hydrolyzed other amides but no longer recognized or catalyzed the hydrolysis of the original substrates. A form of L-fucose isomerase with a lower K_m for D-arabinose was found in the pentose study. In the amidase and pentitol cases, further mutations in regulatory genes were also obtained that significantly altered the induction of the rate-limiting enzyme by the new substrates.

From these studies, it can be inferred that sequential mutations and selections might permit the alteration of substrate specificity of enzymes so that closely related substrates may be metabolized at reasonable rates. A first step that would probably be required in such studies would be mutation in regulatory genes causing over-synthesis of the rate-limiting enzyme. It appears from these studies that after several sequential mutations, recognition of the original substrate may be lost, potentially affecting the overall metabolism of the cell. Hence, if one were setting out to modify an enzyme through sequential mutations and selection, it would be advantageous to work with a gene that has been introduced into a strain in multiple copies, so that the original enzyme activity is not lost.

Another pioneering study that deserves mention is the work of Francis and Hansche (1972), who used chemostats to deliberately modify a specific enzyme, acid phosphatase. By applying intense selective pressure on this growth-rate-limiting enzyme, predicted changes in the enzyme's characteristics were produced. A strain of *Saccharomyces* was used in a culture limited by the supply of phosphate from β-glycerophosphate through the action of a specific acid phosphatase present on the cell surface. The culture conditions were more alkaline than the pH optimum for the enzyme and the concentration of substrate was near the enzyme's K_m. The strain was cultured in a chemostat for approximately 1,000 generations, with the expectation that spontaneous mutants having a superior adaptation to these conditions would arise and take over the culture.

Three such spontaneous mutations were observed. In the first, the overall phosphate-limited growth yield of the strain improved, permitting approximately 30 percent more biomass to form in the chemostat. The second mutation altered the pH optimum of the enzyme to permit a more rapid metabolism of the β-glycerophosphate. The third mutation was an adaptation to the continuous culture apparatus, leading to clumping of cells and reduced washout. The last mutation terminated the experiment. The experiment illustrated the intensive pressures that can be applied to the genes for growth-rate-limiting enzymes through the selection of new substrate-concentration versus specific growth-rate characteristics. Such selections potentially provide a tool for modification of individual enzymes in predictable ways.

All of Francis and Hansche's work was performed without the use of mutagenic agents. By combining general or site-specific mutagenesis and chemostat selection, the frequency of isolation of desired mutants may be increased.

A good deal of basic research remains, however, before such isolations can be routinely performed. One obvious problem, avoided by Francis and Hansche, stems from the necessity for permeation of a substrate into the cell before a particular enzyme may act upon it to produce a flux of carbon and energy leading to determination of growth rate. Novel substrates may require mutations in separate genes to permit the alteration of permeation and of utilization before the growth rate of a strain would be affected. The probability that mutations will be stable when the chemostat's selective pressures are removed also needs to be experimentally evaluated. For example a 2,4,5-trichlorophenoxyacetic acid (2,4,5-T) degrading *Pseudomonas cepacia*, derived as described below, rapidly lost its herbicide-degrading capacity when cultured on a nonselective medium (Kilbane *et al.*, 1982). Nevertheless, this approach appears to be an extremely exciting and potentially fruitful means for directed alteration of particular proteins for the design of superior and useful strains.

The accelerated evolution of a catabolic pathway without predetermination of the enzymatic steps to be altered can also be carried out, as is illustrated by the selection of a 2,4,5-T degrading bacterium (Kellogg *et al.*, 1981; Kilbane *et al.*, 1982; Chatterjee *et al.*, 1982). A mixed culture was established in a chemostat using several carrying substrates. The inoculum included known strains bearing conjugative plasmids encoding various catabolic pathways for aromatic hydrocarbons and a portion of 2,4,5-T-contaminated soil. Periodically the proportion of 2,4,5-T was increased, until the halogenated acid was the sole source of carbon and energy supporting the culture. A single strain of *Pseudomonas cepacia* was eventually isolated from the community that evolved. This strain could use gram-per-liter concentrations of 2,4,5-T as the sole carbon and energy source, but did not maintain itself in soil in the absence of high concentrations of the herbicide. It was also highly unstable in pure liquid culture in the absence of 2,4,5-T, perhaps suggesting the loss of essential plasmid genes. Isolation of this strain illustrates how intensive selective pressures can be used to force the evolution of a new catabolic pathway and it provides a starting strain from which it may be possible to derive more stable and useful pollution control agents.

Summary: DNA Manipulation

The advent of the *in vitro* and *in vivo* recombinant DNA technologies has made virtually any microbial gene available for precise manipulations, especially for rearrangement with any other microbial gene. The main immediate impediment to the use of these methods is a lack of basic knowledge of normal means for gene exchange in microbes thought to be useful for advanced pollution control. More difficult problems concerning gene expression may occur in some cases, but at present this has not been widely experienced. It is believed that a basic characterization of gene exchange can be readily acquired in any interesting bacterial species, making the full range of these new gene manipulations available for strain construction.

A technique offering great promise for developing new methods of pollution control is the directed modification of enzymes through mutation and selection of their genes. A few pioneering studies have shown that individual enzymes can be derived that will accept unnatural substrates, that have altered kinetic properties or that favor alterations in their physio-chemical environment. In principle these methods can be used to extend the substrate specificity of existing enzymes to pollutants with structures related to their natural substrate, and even to construct entire catabolic pathways by linking together the genes for several such altered enzymes. Complementary to such precise gene manipulations, the application of intensive selective pressures on mixed cultures holds promise for the isolation of novel strains, capable of biodegrading recalcitrant substrates, as a result of an acceleration of natural evolutionary processes.

4

Limits to Biodegradation

The purpose of this section is to identify the factors that limit the biotransformation of compounds in the environment. It seems likely that understanding such factors will suggest ways to overcome them using genetic manipulations. Among the factors addressed will be the physical conditions that confront microorganisms, limitations imposed by the concentration of a target compound, cometabolic phenomena, and interactions among microorganisms.

Environmental Factors

The physical environment of a microbe provides a set of elements that must be conducive to its functioning. One may view microbes catalyzing a desirable transformation either as indigenous to an environment (colonizers) or as transients that must tolerate the conditions present (allochthonous organisms) (Winogradskii, 1949). In a parallel fashion the physical characteristics of a microbe's environment are often studied from either of two perspectives, one concerned with the variations of stress tolerable by a microorganism or one describing the features of a particular organism and habitat that permit colonization.

Descriptions of the features essential for colonization of a habitat are predictive in that these features should be present wherever the organism proliferates. It is dangerous, however, to draw general conclusions from many of the studies of the limits of stress tolerable by an organism because such studies often focus on only one facet of the stresses possible in an environment. Such studies may fail to adequately assess the effects on organisms of other environmental factors, especially factors that may act synergistically. Thus, generalizations should be made with caution.

Consider, for instance, studies of crude oil biodegradation showing that the degradation is enhanced with increasing temperature. In most reports, the mechanism of the enhancement is not discerned and could include shifts in the components of a microbial population, changes in catabolic pathways within an organism, increases in enzyme reaction rates, increases in the rate of growth of the microorganisms, increases in the rate of disappearance of the volatile or toxic products of petroleum metabolism, and increases in the availability of the hydrocarbon substrates to the organism due to physical changes of the oil itself, including dispersion and emulsification. In order to generalize the conclusion that higher temperatures increase oil biodegradation one must first define the mechanism so that the validity of the claim can be reliably assessed in new situations, for instance with different crude oils, various physical settings and microbial species, and within a specific range of temperatures.

Most studies of the influence of environment on biodegradation consider only one factor at a time. Such studies must be applied to new situations with caution, because of the many interacting factors that are possible. Two factors sometimes considered together are pH and redox potential. This stems from the frequent involvement of protons in biological oxidation-reduction reactions.

Hambrick et al. (1980) described the influence of pH and redox potential on the extent of mineralization of naphthalene and octadecane in sediments. They measured the release of $^{14}CO_2$ from labelled substrate under 12 conditions made up of 3 pH values and 4 redox potentials. Overall, biodegradation was enhanced by higher redox potential and higher pH, suggesting that aerobic microorganisms in the upper, oxidized levels of sediment possessed the greatest potential to mineralize these reduced substrates.

Unlike most work that focuses on single environmental factors, this study clearly demonstrated the importance of interacting factors. The oxidation of octadecane, for example, was halved from its maximum at pH 8.0 and redox potential 500 mv, by a drop in pH to 5.0. At redox potentials of 240 mv and 0 mv, this pH shift dropped the extent of mineralization to 25 and 0.6 percent of its value at pH 8.0. This illustrates that a study focusing on only one variable at a time might fail to predict the outcome of a biodegradation in a new situation where the study conditions were not exactly met. This caution needs to be kept in mind when considering the following subsection on environmental factors limiting biodegradation.

Background: The Influence of the Physiochemical Environment on Biodegradation

Temperature. The specific effect of temperature on biodegradation has been studied extensively. In principle, biodegradation rates are expected to follow a simple relationship predicted by the Arrhenius equation, but Vela and Ralston (1978) have considered rates of phenol oxidation in an industrial wastewater and found significant departure from this expectation. They reported that at temperatures from 2-10°C, phenol degradation by the bacteria was proportional to the bacterial growth rate. But as the temperature increased from 10° to 24°C, the rate of phenol degradation by the bacteria increased much more than was predicted by the proportion of the growth rate at the lower temperature. At the higher temperature more phenol was metabolized per cell than was needed to support growth, due to "complex relationships which include population size, rates of oxidation, ambient temperature and probably the metabolic pathways used by the organisms involved." While noting the departure from the expected behavior, the specific causes were not discerned.

Various other reports have shown unpredicted increases in crude oil degradation at increased temperatures. These observations could be due to any of the mechanisms previously mentioned. Atlas (1975) believes that the temperature effect is merely the result of the volatilization of toxic materials in the oil. In a similar vein, Tison and Pope (1980) observed that shifts upward from a temperature to which *Pseudomonas fluorescens* or *Escherichia coli* had become adapted would cause an increase in the percentage of glucose mineralized per cell by these bacteria. The increase was greater with larger shifts and they suggested that this was an indication of thermal stress.

Interfaces. Phase interfaces are well known sites of enhanced biodegradative activity. By positioning themselves at these interfaces, many microbes take advantage of the higher concentrations of compounds that can occur there. With liquid-gas interfaces, compounds accumulate in the liquid surface microlayer, especially in lake and ocean waters. Liquid-solid interfaces offer attachment sites for microbes, adsorption sites and ion-exchange sites for microbes and chemicals. Microbial activity at gas-solid interfaces is not well studied; the presence of conditions conducive to microbial life, i.e., moisture, may preclude the existence of bacteria at such interfaces. In any case, the surface of most microbes is hydrated, and often is coated with slime exuded by the cell. These coatings obscure the nature of the gas-solid interface, since the slime may have some of the properties of solids and of liquids.

Liquid-liquid interfaces formed from an oil and water are often encountered in the environment. Biodegradation of the material in the oil phase usually is limited by the solubility of the oil's components in water, the phase that microbes usually inhabit. Most microorganisms cannot readily survive in the oil phase and even if they can, they must obtain water and water-soluble nutrients at the interface. Although *Cladosporium resinae* has been mentioned as an organism that can grow in the oil phase, its growth is limited by the availability of water (Hill and Thomas, 1955). Liu (1980) has shown that the addition of sodium ligninsulfonate to a culture containing PCBs greatly increases the degradation of the latter compounds. The effect is attributed to the ability of the sodium ligninsulfonate to stabilize PCB-water emulsions, increasing the interfacial surface area postulated to be the limting factor in this degradation.

The availability of solid-liquid interfaces in the microenvironment can be very important to microbes and other biocatalysts (Burns, 1979). Enzymes released from bacteria can accumulate in the soil and remain active for a time, presumably adsorbed to clay particles. Pesticides and other hydrophobic substances exhibit a similar adsorption, either to soil components or to microbial and plant surfaces. Such adsorption can tend to either increase or decrease a compound's degradation. When sorbed to a surface readily colonized by bacteria able to attack it, biodegradation of a compound may be enhanced. Sorption to materials that aren't colonized, and that are resistant to biodegradation, may reduce a compound's degradation. Curiously, a surface resistant to colonization and also somewhat resistant to biodegradation is the cell wall complex of many bacteria. Thus, compounds tend to accumulate when sorbed to a population of microbes lacking the metabolic capability to attack them (Wierich and Gerike, 1981). An example is the accumulation of hydrophobic chlorinated hydrocarbon pesticides by various strains of bacteria, some of which may occur in wastewater sludges (Paris et al., 1977).

Bacteria sorbed to surfaces are generally more resistant to stress imposed by rapid changes in the aqueous environment, presumably due to the buffering influence of the solid surface. Certain plasmids are more readily exchanged among immobilized bacteria than among those in the liquid phase (Matney and Achenback, 1962; Lawton et al., 1968; Sanderson et al., 1972). Such genetic exchange may tend to protect the bacteria by enhancing their ability to respond to a changing aqueous environment.

Jannasch and Pritchard (1972), ZoBell (1943), and Heukelekian and Heller (1946) all indicate that the addition of inert solids to a culture can affect the catabolic activity of certain bacteria. They report increases in bacterial numbers with increasing solid surface areas in cultures having low nutrient concentrations. Reversible attachment of bacteria and adsorption of nutrient to the surface appeared to be involved. Novakova (1972a and b) has looked at the effects of clay concentration on glucose biodegradation and observed that non-expanding clays have no influence on degradation rates, whereas lattice clays have a dramatic stimulatory effect. This is presumably due to the ability of the latter type of clay to remove toxic metabolic products and to regulate the pH of the microenvironment.

Other examples indicate that the activity of some bacteria is associated with the type of material to which they attach. Pseudomonads attached to polyethylene, for example, appear to metabolize amino acids faster than either free-living cells or those attached to glass (Fletcher, 1979).

Solid-liquid interfaces can strongly impede the mobility of microbes and thereby limit their access to pollutants in contaminated soils. Simple models suggest that surfaces wetted with a water film of thickness less than the minimum dimensions of a particle will develop capillary forces sufficient to bind the particle to the surface (Tadros, 1980). These conditions are typically found in soils and are among the forces that bind bacteria to surfaces (Wallace, 1978). Such forces may severely restrict the movement of an introduced, engineered microbe into the soil column where its target compound resides. The model implies that the movement of bacteria will be encouraged by flooding of the soil pores, causing the water film around solid surfaces to become many times the maximum dimension of a bacterial cell. Measurements of bacterial penetration into a soil column confirm these expectations (Madsen and Alexander, 1982) showing very restricted movement in the absence of water percolation. These findings suggest that bacterial strains colonizing plant roots (Schroth and Hancock, 1982; Lynch, 1982) may be preferred as vehicles for genetic manipulations designed for in situ treatment of soil pollutants, because these bacteria can penetrate into the soil column along with the plant root.

Modification of solid surfaces by deposition of a glycoprotein film usually precedes the attachment of bacteria. Factors involved in surface conditioning and microbial adhesion have been reviewed (Fletcher *et al.*, 1980) and include the specific structure of the bacterial surface, the nature of the glycoprotein conditioner, the prevailing ionic strength and the composition of the solid surface. At least the bacterial surface and potentially the glycoprotein composition can be manipulated genetically to alter the attachment of an introduced, engineered strain.

Light. Light is an environmental factor of particular importance to photosynthetic microorganisms (Halldal, 1980). The degradative activities of these organisms are not widely appreciated and it is worth recalling that photosynthetic organisms have available to them a large amount of free energy that potentially can be used to drive otherwise favorable biotransformations. Cerniglia *et al.* (1980) have reported that oxidation of naphthalene by cyanobacteria and microalgae can only take place during periods of illumination. In this case the light appeared to be acting as an energy source for the bacteria, but may also have been involved directly in the degradation via photooxidation of the naphthalene. Reich and Bartha (1977) observed that the synergistic action of sunlight and soil microbes resulted in the slow mineralization of a polybutene film mulch. The microbial attack was enhanced in this case by the photodegradation of the film to smaller molecules and chain fragments with polar ends and side groups.

Hydrostatic pressure. Little research has been devoted to the influence of hydrostatic pressure on biodegradation. Jannasch *et al.* (1971) have suggested that the absence of spoilage observed in food contained in a sunken submarine 1540 meters below the ocean surface could be explained by the excessive hydrostatic pressure. Nevertheless, the recent discovery of chemolithotrophic bacteria colonizing deep sea trenches proves that properly adapted microbes can function at enormous pressures (Karl *et al.*, 1980). Obligate barophiles have even been described (Yayanos *et al.*, 1981). The potential capacity of such strains to degrade or to be influenced by organic pollutants of human origin is unknown.

Water activity. Any particular microbial species can tolerate only a limited range of osmotic pressures, whether produced by ionic or non-ionic solutes (Griffin, 1981). The modification of water by such solutes is best expressed in terms of the activity of the water, k, a dimensionless parameter equal to the quotient of the fugacity of the solution and that of pure water (Reid, 1980). Bacterial life is found only in water of activity between 1 and 0.7, with species tolerant of the lower activities known as halophiles or osmophiles. These limitations are well known and widely exploited for the preservation of food, for instance in pickling brines and sugar syrups. In wastewater streams and polluted environments, water activity may be reduced by salts or by solvents, the latter either being miscible with water or forming a separate phase. Solvents can be biocidal both from a reduction in water activity and by directly disrupting microbial membranes. Even when not biocidal, increased salinity adversely affects the degradation of certain compounds in aquatic environments (U.S. Environmental Protection Agency, 1979; Davis *et al.*, 1979).

Bacteria use at least two genetically controlled strategies to tolerate saline environments (Brown, 1976). The first is the synthesis of enzymes that function normally only at elevated ionic strengths. The second is the synthesis of internal solutes, especially proline and glutamate, that raise the internal osmolarity so as to balance that of the external environment, but without interfering with normal metabolic functions. Because the latter adaptation probably involves fewer genes, it appears to be more amenable to alteration by genetic manipulation.

The fungi are generally better able to tolerate environments with low water activity than are bacteria and they may employ additional mechanisms, especially to tolerate direct exposure to non-aqueous phases. Discerning some of these mechanisms may be especially important to potential biological treatments of materials in toxic waste landfills, since these often contain high concentrations of hydrophobic compounds and may exhibit water activities below the minimum for soil bacteria.

pH. Soil bacteria are often quite sensitive to pH and typically will not grow at pH values less than 5 (Gray, 1976). The pH of a soil depends upon many factors, especially the soil's composition as well as the vegetation, climate, and geology of the region in which it occurs. Gradients of pH can easily occur, especially if surface vegetation and the composition of the lower soil horizons produce opposite and extreme pH values. Gray mentions two examples where the decomposition of pine needles at a soil surface generated pH levels in the range of 3.5-4.0 while the underlying horizon was calcareous (pH 8). In this case a pH gradient was created that strongly affected the availability of metal ions and organic acids at various depths in the soil column, leading to a stratification of the microbial populations across the pH gradient.

In addition to gradients of pH with soil depth, microscopic gradients can exist from the surface of a charged particle into the bulk phase. McLaren and Skujins (1968) have demonstrated the validity of the relationship:

$$pH_s = pH_b + 0.325\mu$$

where pH_s is the pH at the surface of a charged particle, pH_b is that of the bulk phase and μ is the mobility of the charged particle in $\mu m\ sec^{-1}\ volt^{-1}\ cm^{-1}$. Since negatively charged clay particles are widespread in soils, the pH of microenvironments for enzymes or microbes attached to them may differ significantly from that of the bulk soil. The pH near the surface of such particles typically is more acidic than the bulk phase, permitting obvious predictions of altered microbial activity to be made if the pH of the bulk phase and the microbe's response to pH are known.

Some environments may exhibit extremely acidic or alkaline pH values which may be beyond those tolerable by typical soil microbes. Bacteria colonizing environments with a pH as low as 0.7 (Fliermans and Brock, 1972) or as high as 11 (Horikoshi and Akiba, 1982) are known. The organisms colonizing these habitats are usually specialists that may even contribute to the maintenance of the extreme pH. So far no biodegradative activity relevant to pollution problems has been found with such microorganisms.

Toxicants. Toxic substances in waste streams and polluted environments present complex and little studied problems impacting the biological treatment of wastes. To approach this problem requires an initial assessment of the nature and abundance of the toxicants. Next, their availability to the microbe must be considered, since many toxicants may be sorbed, partitioned into non-aqueous phases or sequestered by complexation. The degradation of one component of a polluted environment, for instance humic material, could release sequestered toxicants, a potentially suicidal action from the microbes' perspective. Since the abundance, variety and disposition of microbiocides in polluted environments is often unique to each site, this factor is usually ignored until the failure of a normally efficacious microbial treatment occurs.

In terms of designing superior strains, it is useful to note that microbes differ substantially in their tolerance to general chemical biocides such as phenol, hypochlorite and various solvents, but the genetic basis for this tolerance has not been systematically explored. It may also be noted that bacteria employ a number of specific strategies to avoid heavy metal toxicity (Summers and Silver, 1978), some of them known to be specified by genes on plasmid DNA and potentially transmissible to related strains (Clark et al., 1977). An example of the variety of possible specific actions are the Hg-reducing, and the Hg-methylating activities of bacteria. The utility of these activities to environmental microbes is suggested by the ease of isolation of mercury-resistant strains in mercury-contaminated sediments (Walker and Colwell, 1974).

Toxicants may even affect the biodegradation of compounds by resistant species. Copper concentrations have been observed to negatively correlate with the potential for substrate assimilation and mineralization by estuarine bacteria (Goulder et al., 1979). In contrast, PCBs have produced slight stimulations in glucose uptake velocities in laboratory cultures of freshwater bacteria (Fisher et al., 1974) but were inhibitory to nitrifiers (Sayler et al., 1982). In these cases the toxicants impose environmental stresses and their impacts must be assessed in terms of changes in growth rates and metabolism by bacterial populations. These changes alter the overall fluxes of the biodegraded compounds through the ecosystem. These subtle influences on the makeup of microbial populations are often more difficult to assess than overtly lethal effects.

The creation of new compounds through microbial action is another little studied aspect of environmental toxicants. Bartha (1969) has described the formation of azo compounds from the acylanilides, Propanil and Solan, in the soil. Such transformations were suggested to be catalyzed by soil microbes and the products were resistant to further degradation. Hsia et al. (1977) and Sundstrom (1982) found that these products are more toxic than their parent compounds, especially to mammalian cells. The methylation of mercury is another example of a microbial conversion that enhances the toxicity of a pollutant. It must be emphasized, however, that toxicity is a relative quality that can only be assessed with respect to the exposure of a particular organism to a particular concentration of a compound. In the examples of microbial toxification of pollutants mentioned above, the microbe is usually less adversely affected by the product of the transformation than are various higher organisms such as vertebrates. The increase in toxicity mentioned in the scientific literature usually reflects the higher organisms' perspective.

Concentration of Target Compound

A particularly important factor affecting the biotransformation of an environmental compound is its concentration. To be attacked efficiently, a compound must be readily bound by permeases or other proteins that will make it available to intracellular metabolism. Even compounds that partition strongly into microbial cell walls by purely physical means must ultimately be bound by the proteins that initiate metabolism.

Some compounds may be effectively transformed by exoenzymes, perhaps produced by an engineered microbe. Even though transport to the intracellular space is not required in this case, the rate of transformation responds to the substrate concentration according to the same principles that govern compounds requiring intracellular metabolism.

The characteristic relationship between the steady state rate of a compound's transformation by an enzyme, and its concentration, is hyperbolic. In the simplest case it is described by the Michaelis-Menten equation (White *et al.*, 1964). Whether a particular reaction involves only a single substrate and enzyme form, as in the simplest case, or more than one substrate and enzyme form, all equations describing the rate of compound's transformation versus substrate concentration asymptotically approach a maximum transformation rate at high substrate levels. The reaction velocity always decreases in proportion to substrate concentration at low substrate concentrations to a reaction rate of zero at zero substrate. In principle, therefore, the rate of transformation of an exoenzyme's substrate in the environment could be quantitatively described if the applicable rate equation were known for the enzyme operating under defined steady state conditions. In reality, exoenzymes in soils and sludges are usually sorbed to particles, as may be their substrates (Burns, 1979; 1980). Under these conditions, it is difficult to know the substrate's ambient concentration in the vicinity of the enzyme. In addition, the usual steady state rate equations may not apply. For instance, exoenzymes bound to a particle in the space around the microbial colony producing them may not obey the assumptions inherent in the normal steady-state analysis of reaction rates, particularly the assumption that the concentration of enzyme be negligibly less than that of the substrate. Mathematical treatment of enzyme kinetics, when the concentration of enzyme is an appreciable fraction of the concentration of the substrate, has been described (Straus and Goldstein, 1943; Goldstein, 1944). Under these conditions the concentration of free substrate is noticeably less than that of total substrate, since an appreciable fraction of the substrate is bound as a complex with the enzyme. This condition is frequently encountered by microbial enzymes in environments with low nutrient levels, but these descriptions of enzyme reaction rates have not yet been applied to the analysis of such situations.

The hyperbolic kinetics of an enzyme reaction are strikingly similar to the growth rate versus substrate concentration curves for bacterial growth, originally described by Monod (1942). This may not be too surprising because microbial growth under the fixed conditions of an energy-limited, single-substrate, experimental culture which this equation most often reflects, usually involves a particular enzyme reaction that limits the growth rate by limiting the flux of available energy in the cell. Under these conditions the limitation of growth rate at the higher substrate concentrations reflects the saturation of the rate-limiting enzyme's capacity to metabolize its substrate. At low substrate concentrations, however, the parallel between growth rate and enzyme kinetics breaks down. In this range, microbial growth rates differ qualitatively from enzyme reaction rates; while the rate of an enzyme reaction goes to zero when the substrate concentration reaches zero, the growth rate of a microbe reaches zero at a real, non-zero substrate concentration. Just below this concentration negative growth rates occur. One widely used mathematical formulation of this behavior was described by Pirt (1975):

$$\mu = \frac{\mu_m S}{(S + K_s)} - m\ Y_{EG}$$

where

- μ = the specific growth rate (in units of $time^{-1}$)
- K_s = half maximum saturation constant (in units of concentration)
- S = substrate concentration
- m = maintenance coefficient (a constant in any given habitat)
- μ_m = the maximum specific growth rate
- Y_{EG} = the growth yield if there were no maintenance requirement, i.e., the theoretical maximum possible growth yield from the energy source (a constant for any given microbe and habitat)

As the equation illustrates, a certain amount of growth substrate, m, is needed simply to maintain an organism in a state in which it is healthy and can respond to changes in its environment. This "maintenance ration" is required for functions like the generation of the membrane potential and the replacement of damaged macromolecules. Thus, the maintenance ration may vary for the same organism from environment to environment, for instance at different pH values and ionic strengths, in the presence of toxicants, etc. An organism must do different amounts of work to maintain itself in these various conditions. It is anticipated that the more stressful the environment, the higher the maintenance ration exhibited by an organism, a likelihood of potential importance for engineered microbes designed to mineralize low concentrations of substrates in unfavorable habitats.

The microbial growth rate equation is helpful for understanding certain limits that the concentration of a target compound may impose on the design of a microbe intended to catalyze its biodegradation. First, it is usually desirable for the target compound to serve as an energy source for the microbe since energy sources are typically metabolized by enzyme systems that are abundant in a cell and that lead to rapid and complete degradation of the compound. It is also often easier to manipulate strains in the laboratory through their energy sources. Cultures are easily set up that permit the growth only of mutants or constructed strains utilizing novel or recalcitrant energy substrates. The mathematical description of growth rate given above convincingly applies to energy-limited cultures, although, in principle, it can be applied to any growth-limiting substrate, such as a required inorganic element or a vitamin.

If a target pollutant serves as an organism's predominant energy source in the environment, then the equation readily points out the minimum concentration that the organism can dissimilate efficiently. At concentrations between K_s and the maintenance ration, an organism will dissimilate its growth substrate very much more slowly than its genetic potential could allow, and in this sense it is inefficient. If one were to attempt to modify an organism to enhance the rate of degradation of an environmental compound used as its principal energy source, it would be important to be aware of these relationships so that the organism would be designed for an appropriate growth response at environmentally relevant concentrations of substrate. Yet these implications have often been ignored by microbial physiologists and biochemists. For instance, typical enrichment cultures for non-toxic compounds employ concentrations of substrate on the order of 0.1 percent or greater. This often permits isolation of strains that are convenient to work with, i.e., ones giving overnight cultures of high density in simple single-substrate medium. Most such strains are characterized by relatively high μ_{max} and K_s and frequently show strong catabolite repression by more readily used substrates such as glucose or succinate. These same strains would probably be found to dissimilate the substrate on which they were selected only very slowly at concentrations of a few ppm or less. Yet typical environmental pollutants occur at ppm or lower concentrations.

Until recently, only marine microbiologists have been concerned with organisms colonizing environments with low nutrient concentrations. In contrast to those described previously, this class of microorganisms has a relatively high affinity for substrate (low K_s) but often exhibit a low maximal growth rate (μ_{max}).

Two types of microbes, the slow growing, high affinity, and the rapidly growing, low affinity growers, have been classified according to their strategies for survival under various conditions. Winogradskii (1949) was the first to notice them and called them the autochthonous and zymogenous populations, respectively. More recently the terms R- and k- strategist have been applied as well as the description oligotroph and eutroph (Jannasch, 1979). Poindexter coined the name "copiotroph" to replace eutroph, and to draw a clearer distinction to the term oligotroph (Poindexter, 1981a). Copiotrophs have the capacity to grow very rapidly in the presence of high levels of utilizable substrate but become dormant when levels are low. Some copiotrophs have very stable resting states which allow them to persist for long periods in the absence of substrate. The copiotroph, therefore, can compete with other microbes in its environment by outgrowing them during periods of plenty and by being resistant to death during periods of famine. Microbes isolated from typical enrichments described above are usually copiotrophs.

The oligotroph, by contrast, is an organism that grows slowly, but efficiently. It has a high affinity for utilizable substrates and grows continually in low concentrations of nutrient, provided that the environment provides it with a sufficient flux of nutrients (Poindexter, 1981b). Such organisms are unable to exploit surfeits of nutrients by a rapid growth response, but they are capable of the deposition of reserve materials during periods of plenty. An additional property of the oligotroph is a high likelihood of attachment to solid surfaces (Hirsh et al., 1979). Veldkamp and Jannasch (1972) have demonstrated the relative competitiveness of oligo- and copiotrophs in chemostats. They showed the predicted advantage to oligotrophs at low concentrations of nutrient and to copiotrophs at high concentrations of nutrient, using organisms growing on lactate.

Most microbes in the environment, however, do not grow on single growth substrates. They have many potential growth substrates available, although particular habitats may provide one or a few substrates in abundance, for instance, wastes from the dairy and some chemical industries. When a multitude of growth substrates are present, the concentration of individual components is often low. Microbes inhabiting such environments, e.g., soils, are viewed as in a dormant or near dormant state, limited by the available energy sources (Gray, 1976). Microbes in such environments compete intensively for the energy available and may utilize several compounds simultaneously, as well as incorporating exogenous vitamins and amino acids that may be present, to spare the energy otherwise needed for biosynthesis (Harder and Dijkhuizen, 1982). Oligotrophs would be expected to be successful in such environments because of their greater affinity for utilizable substrates, and presumably because they possess the other attributes mentioned. The organisms dominant in eutrophic environments, in contrast, will exhibit a rapid growth response supported by high concentrations of a growth substrate, and will tend to specialize their uptake of substrates, concentrating on the most readily dissimilated compounds. For instance, enteric bacteria typically exhibit diauxic utilization of various sugars, i.e., sequential utilization of sugars when high levels of several sugars are available. Just as microbiologists may ignore oligotrophs when isolating strains from enrichments, so might molecular biologists forget that diauxie is usually exhibited only in the presence of an abundance of readily utilizable substrates (Harder and Dijkhuizen, 1982). This is rarely a characteristic of polluted environments. The typical single substrate, energy-limited growth rate curve described previously must therefore be reconsidered and interpreted as representing the overall concentration of all available energy substrates in a microbe's environment, if it is to describe microbial growth in typical polluted environments. It must also be realized that in typical situations the target compound(s) in a polluted environment will make up only a fraction of the compounds supporting bacterial growth.

Background: Oligotrophic Bacteria

Tempest and Neijssel (1978) have pointed out a couple of metabolic strategies that may be used by oligotrophs to increase their metabolic efficiency. They suggest that transport proteins with high affinity for nutrients will be constitutively produced, but that the actual compound-transforming enzymes will be repressed unless a utilizable substrate is available to the cell. A corollary of this is that oligotrophs will carefully modulate uptake of non-nutrients when these are present in excess of the carbon- or energy-supplying substrates. Presumably, energy metabolism will be highly efficient and tightly coupled to a simplified set of biosynthetic requirements so that the highest possible growth yield will be obtained from available compounds. Finally, there must be careful coordination within the organism of the relative rates of macromolecular synthesis so that growth occurs in a balanced fashion. While not all of these modifications have been demonstrated in bacteria colonizing low nutrient habitats, Tempest and Neijssel (1978) found examples supporting the possibility of these kinds of modifications, including a strain of *Saccharomyces cerevisiae* that oversynthesized glutamate dehydrogenase, an enzyme with a relatively low affinity for its substrate. The oversynthesis compensated for the low affinity of the enzyme for its substrate by depleting the internal pool of ammonia within the cell as enzyme-substrate complex, permitting a sufficient nitrogen flux to support growth. They also described a *Klebsiella aerogenes* that elaborated both a high- and a low-affinity glycerol assimilation pathway in response to high versus low concentrations of glycerol in its environment. The extent to which these particular modifications are found in typical oligotrophs is not known.

Particular morphological adaptations apparently also occur in response to low nutrient concentrations. Kuenen *et al.* (1977) have suggested that microbes growing at low concentrations of nutrients will typically show high surface to volume ratios. The fact that prosthecate and stalked bacteria are commonly found in environments characteristically low in nutrients would seem to bear this out. Some of these bacteria undergo dramatic changes in cell shape in response to nutrient limitation. Perhaps the best example of this is the response of a species of *Ancalomicrobium* (Dow and Whittenbury, 1980). In high nutrient concentrations, its cell shape can be either coryneform, rod or coccobacillar. At lower concentrations, however, the cells are somewhat rod-shaped and possess multiple long prosthecae. In another example Larson and Pate (1976) have shown that lengthened prosthacae enhance the rate of nutrient uptake by *Anticcacaulis biprosthecum*.

Application of Oligotrophs for Pollution Abatement

Many toxic and persistent *in situ* pollutants, for instance TCDD in the soils of Seveso, Italy and Vietnam, are present in the environment in concentrations too low to contribute to the growth of typical copiotrophs. When the molecular details of the strategies used by oligotrophs to cope with low nutrient concentrations are better understood, it may be possible to envision the use of these strategies to attack such pollutants. To develop organisms in this way, work must first be undertaken to explore the limits of oligotrophs' abilities to utilize dilute nutrients and to identify the genes responsible. Modern gene manipulations should permit the transplantation of the genes for enzyme functions found in oligotrophs, including transport proteins with high affinity for substrate, into organisms that exhibit a more rapid maximal growth rate on the same substrate. Further manipulation of the recombinant organism, such as the amplification of the number of copies of such genes, may overcome any inherent limitations of these enzymes, such as a combination of low K_m with a low V_{max}. The resulting oversynthesis of such proteins might permit a carbon flux suffient for the reasonably rapid dissimilation of the target substrate by the recombinant organism. Although speculative, this is but one example of an approach that appears to offer some new opportunities.

Co-metabolism

Since Leadbetter and Foster's (1959) discovery of the oxidation of ethane, propane and butane by the methane-degrading bacterium *Pseudomonas methanica*, the terms co-metabolism or cooxidation have been used often in the scientific literature. Over the years, however, a certain amount of confusion, if not controversy, has accumulated regarding reports of this phenomenon. For example, the terms cooxidation and co-metabolism are considered to be interchangeable by some authors but are defined in more restrictive terms by others. This confusion has been recognized and Dalton and Stirling (1982) have recently proposed a revision in the definition of co-metabolism to clarify the use of the term. They define co-metabolism as, "the transformation of a non-growth substrate in the obligate presence of a growth substrate or another transformable compound." In this definition, a non-growth substrate is defined as "a compound that cannot support cellular division, as opposed to producing an increase in biomass." This permits the notion that a non-growth substate could be incorporated into cellular components or otherwise contribute to metabolism, although in a non-essential way. Past definitions requiring that co-substrates not contribute energy to cellular metabolism led to some of the earlier confusion. Dalton and Stirling go on to relegate metabolisms of non-growth substrates in the absence of another substrate, for instance metabolisms by resting cell suspensions, to "fortuitous metabolism, oxidation, dehalogenation, etc." They make an exception for resting cells whose co-metabolism is *dependent* upon dissimilation of a metabolic reserve, for instance poly-β-hydroxybutyrate, which serves as the "other" substrate. Their definition avoids the inherent difficulties in other definitions while encompassing or clearly separating most previous reports of co-metabolism and retaining a useful term rather than abandoning it as recommended by Hulbert and Krawiec (1977).

Much of the historical confusion concerning co-metabolism stems from a lack of a clear notion of the underlying mechanisms. There can be several different underlying reasons for the transformation of a non-growth substrate only in the presence of another compound. Among the possible mechanisms are situations in which: 1) the non-growth substrate is unable to act as an inducer of the pathway(s) needed for its own transport into the cell or metabolism, but the other substrate does; 2) the structure of the non-growth substrate prevents its metabolites from acting as nutrients; 3) the non-growth substrate or its metabolites are co-repressors for a growth-limiting cellular function, and the other substrate relieves the repression; 4) growth on the other substrate may provide energy needed for the co-metabolism; or 5) in some poorly designed experiments, metabolism at the expense of the growth substrate may simply raise the total biomass in the system to a point where uptake and transformation of the co-substrate become noticeable.

While there is some controversy as to whether co-metabolism occurs in nature or not, it remains a plausible approach to the *in situ* treatment of pollutants. If an organism were designed to utilize a particular carbon and energy source (growth substrate) as well as to carry out a particular transformation of a target pollutant (non-growth substrate), the application of large numbers of that organism together with the growth substrate could possibly lead to the effective transformation of the target co-substrate. Alternatively, basic research on identified co-metabolisms may disclose underlying mechanisms such as (1) and (3) above, which might then be manipulated by genetic means, to create a constitutive mutant able to act directly without the growth substrate, i.e., creation of a strain able to directly metabolize the target compound from a parental strain exhibiting co-metabolism of the target compound.

There are a great many reports of co-metabolic phenomena; Table 4 summarizes many of these (and includes reports of some metabolisms that would probably be excluded from co-metabolism by Dalton and Stirling's definition, if examined closely).

It has been said that the possibilities for exploration of co-metabolism is "limited only by the investigator's imagination" (Horvath, 1972). This seems particularly true today regarding the biotransformation of pollutants. Current understandings of enzyme evolution, adventitious reactions of enzymes, modern mutation and selection methods and ability to reassort and multiply bacterial genes, have opened up many new avenues for the exploitation of co-metabolism.

For example, when a desirable co-metabolic attack on a pollutant is not known, it should be possible to choose an organisms possessing an enzyme attacking a structurally related compound and derive a mutant from it possessing an enzyme attacking the pollutant. The technique of site-directed mutagenesis seems particularly appropriate for this approach. Such mutant genes, once obtained, can be amplified in that or another more desirable host organism so that a suitable pollution control agent is created. This approach to the creation of microbes catalyzing single co-metabolic transformations of target pollutants appears to be highly feasible using methods currently available.

Table 4: Examples of Co-Metabolism

Organism	Co-Substrate	Growth Substrate	Product	Reference
Achromobacter sp.	Amitraz*	yeast extract	not identified	Baker & Woods, 1977
Arthrobacter sp.	m-chlorobenzoate	benzoate	4-chlorocatechol	Horvath & Alexander, 1970b
Bacillus spp.	various n-alkanes, e.g., tetradecane	glucose + nutrient medium	various ketones and diols	Kachholz & Rehm, 1978; 1980
Brevibacterium sp.	2,3,6-trichlorobenzoate	benzoate	3,5-dichlorocatechol	Horvath, 1971
Methylomonas methanica	ethane, propane, butane	methane	alcohols, ketones and acids	Foster, 1962
Mixed culture	cyclohexane	hexadecane	increased mineralization	Beam & Perry, 1974
Mycobacterium rodochrous	cyclohexane	resting cells grown on propane	cyclohexanone	Beam & Perry, 1973
Mycobacterium smegmatis	methane, ethane, ethylene or propylene	resting cells grown on propane	not identified	Lukins & Foster, 1963
Mycobacterium vaccae	cyclohexane	propane	cyclohexanone	Beam & Perry, 1974; Perry, 1979
Nocardia corallina	toluene	hexadecane	2,3-dihydroxybenzoate; α-methylmuconate	Jamison *et al*, 1969; 1971
Nocardia salmonicolor	p-xylene	hexadecane	2,3-dihydroxytoluate; p-toluate	Raymond *et al*, 1967
Nocardia salmonicolor	various β-dimethylnaphthalenes	hexadecane	various β-carboxy methylnaphthalenes	Raymond *et al*, 1967
Nocardia salmonicolor	ethylbenzene	hexadecane	phenylacetate	Davis & Raymond, 1961
Nocardia salmonicolor	p-cymene	hexadecane	cumate	Davis & Raymond, 1961
Nocardia salmonicolor	various alkylbenzenes	hexadecane	carboxylic acids	Davis & Raymond, 1961
Nocardia sp.	4-(methylmercaptan)-phenol	yeast extract	2-hydroxy-5-methylmercaptan muconic semialdehyde	Englehardt *et al*, 1977

(continued)

Table 4: (continued)

Organism	Co-Substrate	Growth Substrate	Product	Reference
Nocardia sp.	4-(methylsulfinyl)-phenol	yeast extract	2-hydroxy-5-methylsulfinyl muconic semialdehyde	Englehardt *et al*, 1977
Pseudomonas aeruginosa	Kepone**	yeast extract	monohydro-kepone	Orndorff & Colwell, 1980
Pseudomonas putida	2-chloronaphthalene	naphthalene	chloro-2-hydroxy-6-oxo-hexadieneoate	Morris & Barnsley, 1982
Pseudomonas sp.	Amitraz	yeast extract	not identified	Baker & Woods, 1977
Pseudomonas sp.	3-chloropropionate	yeast extract	not identified	Slater *et al*, 1979
Pseudomonas sp. B13	isomeric cresols	resting cells grown on phenol	methyl catechols	Knackmuss & Hellwig, 1978
Pseudomonas sp. B13	various chlorophenols	resting cells grown on phenol	various chlorocatechols	Knackmuss & Hellwig, 1978
Soil organism	Diazinon***	root exudates	increased mineralization	Baker & Woods, 1977
Soil organism	Parathion†	root exudates	increased mineralization	Hsu & Bartha, 1979
Soil organism	Malathion††	n-heptadecane	increased mineralization	Merkel & Perry, 1977
Unidentified micro-organisms	Lindane†††	glucose	mineralization	Lyons & Savage, 1979

*Amitraz = 1,5-di(2,4-dimethylphenyl)-3-methyl-1,3,5-triazapenta-1,4-diene.

**Kepone = decachlorooctahydro-1,3,4-metheno-2H-cyclobuta[cd] pentalen-2-one.

***Diazinon = 0,0-diethyl-0-[2-isopropyl-4-methyl-6-pyrimidyl] phosphorothioate.

†Parathion = 0,0-diethyl-0-p-nitrophenyl phosphorothioate.

††Malathion = S-[1,2-dicarbethoxyethyl]-0,0-dimethyldithiophosphate.

†††Lindane = 1,2,3,4,5,6-hexachlorocyclohexane, γ-isomer.

Microbial Consortia

Enrichment cultures utilizing single energy sources frequently isolate a community of interacting microorganisms rather than a single strain catalyzing the selected biodegradation. A few such communities have been analyzed in detail. Table 5 mentions a number of examples in which the community is able to catalyze a particular biodegradation better than its component members.

The basis of the interaction in these studies usually involves the partial transformation of the enrichment substrate by one microbial group with subsequent utilization of the product by a second group. The second group, in many cases, excretes some factor essential to the first group or removes the product of the first group's metabolism, when this is inhibitory. Thus, there is a mutual positive feedback between these two groups. Often there is a third tier of organisms stably maintained in such cultures. These organisms utilize excreta from the first or second group, but are not essential to the degradation of the substrate. As defined in the background statement, these associations are mutualistic or commensalistic.

The term consortium has been applied to microbial interactions such as those described in Table 5. As defined by Brock (1979), "a consortium is a two-membered culture or natural assemblage in which each organism benefits from the other." Historically the term consortium referred to very close associations of pairs of organisms interacting metabolically, as in the "species" *Methanobacillus omelianskii* and *Chloropseudomonas ethylica* (Bryant *et al.*, 1967; Gray *et al.*, 1973). In each of these examples, the apparent species actually consisted of two interacting organisms that were mistaken for a single organism.

The importance of biodegradative interactions among microorganisms in nature is not clearly established, except for interactions of H_2-producers and methanogens in anaerobic environments. Interactive communities obtained from elective culture are not necessarily representative of the situation in the environment. They could have arisen during isolation in response to the intense selective pressures applied. Slater and Bull (1982) have made the interesting suggestion that interactive cultures like these may be a common precursor to the emergence of a single strain capable of the same transformation. Such strains would presumably arise by natural gene exchange among the key members of the consortium. They imply that the isolation of individual strains may frequently be the result of microbial evolution in the enrichment culture. This notion is in accord with Chakrabarty's experiments with plasmid-assisted breeding, in which organisms from polluted environments were cultured on 2,4,5-T in a chemostat, along with laboratory strains harboring conjugative plasmids (Kellogg *et al.*, 1981). In this work, a single 2,4,5-T degrading pseudomonad was eventually isolated, although a consortium appeared to precede it (Kilbane *et al.*, 1982; Chatterjee *et al.*, 1982).

Table 5
Biodegradation Enhanced by the Interaction of Microbes

Organisms	Comments	Reference
Hydrogenomonas sp. and *Arthrobacter* sp.	*A.* grew on p-chlorophenylacetic acid produced from DDT[a] by *H.*	Pfaender and Alexander, 1972
Pseudomonas sp. and *Achromobacter* sp.	The pair grew on Silvex[a], but not the individual strains. Chloride, CO_2 and a small amount of 2,4,5-trichlorophenol was produced.	Ou and Sikka, 1977
Bacillus polymyxa and *Proteus vulgaris*	*P.* produced nicotinic acid required by *B*; *B.* produced biotin required by *P.*	Yeoh *et al.*, 1968
Chlorobium limicola and *Desulfovibrio* sp.	*C.* used HS^- as electron donor producing $SO_4^=$; *D.* used $SO_4^=$ as electron acceptor, producing HS^-.	Gray *et al.*, 1973
Saccharomyces cerevisiae and *Lactobacillus casei*	*S.* produced riboflavin required by *L.*	Megee *et al.*, 1972
Nitrosomonas sp., *Nocardia atlantica* and *Pseudomonas* sp.	The conversion of NH_4^+ to NO_2^- by *Nit.* increased in the presence of *Noc.* and *P.*	Jones and Hood, 1980
Lactobacillus plantarum and *Streptococcus faecalis*	*L.* produced folic acid required by *S.*; *S.* produced phenylalanine required by *L.*	Harrison, 1978
Pseudomonas sp., unidentified bacterium, *Trichoderma viride*, *Ps. putida*, budding yeast, *Flavobacterium* sp. and unidentified pseudomonad	The first four organisms were primary degraders of Dalapon[a]. The others grew on the waste products of the first four.	Senior *et al.*, 1976
Methanobacterium sp. and "S" organism	"S" fermented ethanol to H_2; *M.* used H_2 for growth and methane formation.	Bryant *et al.*, 1967

Table 5 (continued)

Organisms	Comments	Reference
Methanobacterium sp. and *Desulfovibrio desulfuricans*	*D.* released acetate which was fermented by *M.* to produce methane.	Cappenberg, 1975
Pseudomonas putida and *Pseudomonas* sp.	The mixed culture used polyvinyl alcohol as sole carbon source, the isolated strains could not. *Ps.* sp. degraded PVA but required growth factors from *P. put.*; *P. put.* grew on PVA metabolites from *Ps.* sp.	Sakazawa *et al.*, 1981
Three-pseudomonads and *Hyphomicrobium* sp.	The pseudomonads oxidized methane; *H.* used the methanol, preventing its accumulation to levels that inhibit the pseudomonads.	Wilkinson *et al.*, 1974

See Slater and Bull (1982) for citations of 29 other organic compounds degraded by the combined activities of 2 or more microbial populations.

---

[a] DDT = 1,1,1-Trichloro-2,2-bis[*p*-chlorophenyl]ethane
Silvex = 2-[2,4,5-Trichlorophenoxy]propionic acid
Dalapon = 2,2-Dichloropropionic acid

Similarly, the analysis of a community degrading 2-chloropropionamide revealed a stable community based on a co-metabolic deamination (Slater and Bull, 1982; Slater et al., 1979). One member of the community deaminated the primary substrate to 2-chloropropionic acid, a substrate used by another community member. After prolonged culture of the community, the latter acquired the ability to attack the primary substrate. Although the possibility that this represented the development of a new enzymatic activity was not ruled out, it could be explained as the result of a transfer of genes from the member of the consortium initially catalyzing the deamination of the primary substrate. The new recombinant strain became more numerous than its immediate predecessor, but it did not completely displace the other members of the community.

Background: Overview of Microbial Interactions

As illustrated above, interactions between microbial populations can affect the biodegradative capacity of the community as a whole. Slater and Bull (1978) have classified pairwise microbial interactions on the basis of whether a particular member is benefited (+), inhibited (-), or unaffected (0) by the presence of another. Six types of pairwise interactions are possible: 1) neutralism (00), 2) amensalism (0- or -0), 3) commensalism (+0 or 0+), 4) predation (+- or -+), 5) mutualism (++), and 6) competition (--). Their article offers illustrations from the primary literature of examples of each of these kinds of interactions. For the purposes of genetic engineering, it seems sufficient to be aware that interactions of this sort are possible whenever an organism is released to a polluted environment. These interactions may impede the treatment being attempted by limiting the proliferation of the introduced strain, may improve the treatment if the introduced strain is benefited by another microbe or may leave the strain unaffected. It usually is not possible to predict in advance the detailed impact of such interactions on introduced microbes since it is unlikely that a sufficiently detailed knowledge of the indigenous microflora would be available (Alexander, 1971).

Three kinds of interactions invite qualitative predictions concerning the biodegradation of target compounds. First, mutualist interactions offer the possibility to treat some relatively recalcitrant wastes, as has been implied in the foregoing discussion of microbial consortia. Seeding a polluted site with the members of a consortium previously selected for the degradation of a particular target compound should bring about a desirable result, provided that each member survives the physical stresses of the habitat and that each has access to the pollutant and can operate on the concentration of the target compound found at the site.

Second, any introduced microbe may be subjected to predatory grazing by other organisms, especially protozoa. Habte and Alexander (1977, 1978) and Alexander (1981) have proposed that the grazing of protozoa on bacteria is a major factor regulating bacterial populations in the soil. Gude (1979) has performed experiments that suggest that protozoan predation on bacteria acts to select certain types of bacteria in activated sludge. Alexander (1981) has recently discussed the interactions of microbial predators and their prey pointing to the influence of population densities and the energy demand for finding and attacking the prey as key variables. Thus, bacterial populations below 10^6 organisms per g of soil typically escape serious predation, while densities above this may engender increases in the predator population in a reciprocal fashion.

The third predictive interaction is competition. Any organism introduced into a non-sterile environment is highly likely to experience competition for those factors essential to its proliferation. These factors are likely to include energy sources, essential elements, space, and attachment sites. The relative competitiveness of pairs of microbes with differing affinities for a growth substrate and differing growth rate potential, have been modeled mathematically (Slater and Bull, 1978). These models illustrate the importance of matching an organism's affinity for its target compound with the compound's environmental concentration.

Dehalogenation of Environmental Pollutants

Halogenated compounds are frequently found among the toxic and persistent environmental pollutants. Except in marine environments, natural halogenated compounds are scarce and generally considered inimical to life. The scarcity of natural halogenated compounds is reflected by a paucity of enzymatic means to break carbon-halogen bonds, and a general tendency to environmental persistence. The lipophilicity of many man-made organohalogens may also exacerbate their biodegradative recalcitrance by leading to sequestration in non-aqueous phases. It certainly exacerbates the toxic potential of these compounds by causing them to concentrate in the upper trophic levels of food webs (Tinsley, 1979). Nevertheless, as illustrated in Table 6, halogenated pollutants can be biodegraded, although they may not often be mineralized. A discussion of dehalogenation will illustrate the impacts of the factors mentioned in this Section.

Biological dehalogenation of many simple organohalogens is likely to fail when these compounds occur in very concentrated or very dilute form. If too concentrated, for instance solvents in chemical landfills, the compound may either reduce the activity of water to a level below that permitting microbial life, or it may form a separate biocidal non-aqueous phase. If too dilute, for instance TCDD in contaminated soils, the compound's concentration may be so low that it is effectively inaccessible by most microbes. Or, the compound may dissolve in the membrane phospholipid bilayer of microbes unable to metabolize it, for instance PCB's in wastewater sludge, thereby decreasing its availability to organisms able to metabolize it. In any of these situations, the degradation can be compromised by other toxic compounds that occur at the same site and which kill organisms or effectively inhibit the relevant biodegradation.

Putting aside physical factors, the degradation of organohalogen compounds proceeds by only a few known biological mechanisms. There appear to be two types of enzymes that can break the carbon-halogen bond, specific enzymes whose primary function appears to be dehalogenation, and a broad group of apparently fortuitous dehalogenating enzymes. The first group is

Table 6
Examples of Microbial Dehalogenations

Organisms	Substrate	Comments	Reference
Achromobacter sp.	2,4-D[a]	inducible oxidation, qualitative release of chloride	Bell, 1957
Acrostalagumus sp.	chloro-substituted aliphatic acids		Jensen, 1959
Aerobacter aerogenes	DDT[a]	reduced cytochrome oxidase required	Wedemeyer, 1966 Wedemeyer, 1967 Plimmer *et al.*, 1968
Alcaligenes eutrophus B9	2-fluorobenzoate		Engesser *et al.*, 1980
Agrobacterium sp.	chloro-substituted aliphatic acids	dehalogenase activity	Jensen, 1960 Jensen, 1957
Arthrobacter sp.	chloro-substituted aliphatic acids	dehalogenase activity	Jensen, 1960 Jensen, 1957
	pentachlorophenol	mineralization	Stanlake and Finn, 1982
Azotobacter sp.	3-chlorobenzoate		Walker and Harris, 1970
Bacillus coli	BHC[a]	anaerobic degradation to monochlorobenzene	Allan, 1955
Brevibacterium sp.	2,3,6-trichlorobenzoate, *m*-chlorocatechol and 3,5-dichlorocatechol	converted to 3,5-dichlorocatechol, 4-chlorocatechol and 2-hydroxy-3,5-dichloromuconic semialdehyde, respectively	Horvath, 1971 Horvath and Alexander, 1970a
Clonostachys sp.	chloro-substituted aliphatic acids		Jensen, 1959

Table 6 (continued)

Organisms	Substrate	Comments	Reference
Clostridium sp.	Lindane[a]	anaerobic degradation to unidentified product	MacRae *et al.*, 1969
Clostridium sporogenes	BHC	anaerobic degradation to monochlorobenzene	Allan, 1955
Escherichia coli	DDT	converted by cell membranes to TDE[a]	French and Hoopingarner, 1970
"Fleischmann's Yeast"	DDT	converted to DDD[a]	Kallman and Andrews, 1963
Hydrogenomonas sp. and *Monilia*	Chlorosuccinic, 3-chlorobutyric, 4-chlorobutyric acid and DDT		Focht, 1971 Focht, 1972
Mucor alternans	DDT	anaerobic degradation to water-soluble products	Juengst and Alexander, 1976
Moraxella sp. B	monohaloacetate	plasmid-encoded haloacetate halidohydrolase	Kawasaki *et al.*, 1981c
Norcardia erythropolis	DDT	converted to DDD	Chacko *et al.*, 1966
Proteus vulgaris	DDT	converted to DDD	Barder *et al.*, 1965
Pseudomonas sp.	1,9-dichlorononane, 1-chloroheptane and 6-bromohexanoate	dehalogenation may be inducible	Jensen, 1957
Pseudomonas sp. B13	2-fluorobenzoate, 3-chlorobenzoate		Engesser *et al.*, 1980 Reineke *et al.*, 1982

Table 6 (continued)

Organisms	Substrate	Comments	Reference
Pseudomonas dehalogens	chloro-substituted aliphatic acids	dehalogenase activity	Jensen, 1960
Pseudomonas putida	Dalapon[a] and others	dehalogenase activity; evolution of an extant dehalogenase	Slater *et al*., 1979 Senior *et al*., 1976
Pseudomonas putida	various 2-haloaliphatic acids		Motosugi *et al*., 1982
Pseudomonas and *Achromobacter*	Silvex[a]	chloride and CO_2 produced	Ou and Sikka, 1977
Ruminal fluid	*o,p*'-DDE		Fries *et al*., 1969
Ruminal fluid	Erbon[a]	converted to 2-(2,4,5-trichlorophenoxy)-ethanol anaerobically	Wright *et al*., 1970
Ruminal fluid	Methazol[a]	converted to 1-(3,4-dichlorophenyl-3-methyl) anaerobically	Gutenmann *et al*., 1972
Sediment and sewage sludge	halobenzoates	anaerobic dehalogenation	Suflita *et al*., 1982
Sewage or its sludge	1,9-dichlorononane, 1-chloroheptane and 6-bromohexanoate		Omori and Alexander, 1978
Sewage or its sludge	Mirex[a]	converted to 1,2,3,4,5,6,7,8,9,10-undecachloro-pentacyclodecane	Andrade and Carlson, 1975

Table 6 (continued)

Organisms	Substrate	Comments	Reference
Soil microorganism(s)	Heptachlor and Heptachlor epoxide[a]	converted to chlordene and 1-exohydroxychlordene, respectively	Miles *et al.*, 1971
Soil-water culture	vicinal bromides (ethylene dibromide, 1,2-dibromochloropropane and 2,3-dibromobutane	converted to ethylene, propanol and butenes, respectively	Castro and Belser, 1968
Streptomyces aureofaciens	DDT	converted to DDD	Chacko *et al.*, 1966
Streptomyces viridochromogenes	DDT	converted to DDD	Chacko *et al.*, 1966
Trichoderma viride	chloro-substituted aliphatic acids	chloride ions liberated	Jensen, 1959 Jensen, 1957 Matsumura and Boush, 1968

[a]2,4-D = dichlorophenoxyacetic acid; DDT = 1,1,1-trichloro-2,2-bis[*p*-chlorophenyl] ethane; BHC = 1,2,3,4,5,6-hexachlorocyclohexane, all isomers; Lindane = 1,2,3,4,5,6-hexachlorocyclohexane, γ-isomer; DDD = TDE = 1,1-Bis[*p*-chlorophenyl]-2,2-dichloroethane; Dalapon = 2,2-dichloropropionic acid; Silvex = 2-[2,4,5-trichlorophenoxy] propionic acid; *o,p*'-DDE = 1,1-dichloro-2[*p*-chlorophenyl]-2[*o*-chlorophenyl] ethylene; Erbon = 2-[2,4,5-trichlorophenoxy]-ethyl-2,2-dichloropropionate; Methazol = 2-(3,4-dichlorophenyl)-4-methyl-1,2,4-oxadiazolidine-3,5-dione; Mirex = dodecachlorooctahydro-1,3,4-metheno-1-H-cyclobuta[cd]pentalene; Heptachlor = 1,4,5,6,7,8,9-heptachloro-3a-4,7,7a-tetrahydro-4,7-methanoindene.

typified by the hydrohalidases, some of which have recently been found to be plasmid-specified (Kawasaki et al., 1981a,b,c). Hydrohalidases (Goldman, 1965, 1972) typically catalyze the displacement of particular halides on mono- and dihaloalkanes, and some other closely related acids. Other hydrohalidases attack fluorine or chlorine and bromine, as well as showing a specificity for the organic acids attacked. Dehydrohalidases, for instance, the enzymes in insects that remove the elements of HCl from DDT, are a second set of specific dehalogenating enzymes.

The group of enzymes catalyzing fortuitous dehalogenations are all normally active in intermediary metabolism, but are also able to accept halogenated substrates. Thus, such enzymes have been termed "adventitious dehalogenases" by Goldman (1972). Most often the reaction that they catalyze generates a chemically unstable form of the halogen which spontaneously decomposes. An example is the halohydrin, 2-fluoromalate, produced from fluorofumarate by fumarase. This compound decomposes to produce fluoride and oxaloacetate (Clarke et al., 1968). Other examples include the halohydrin made from 4-chloro-cis, cis-muconic acid by lactonizing enzyme during the metabolism of 2,4-dichlorophenoxyacetic acid (Bollag et al., 1968; Tiedje et al., 1969) and the dioxygenase that removes fluoride from o-fluorobenzoate, producing catechol (Goldman, 1969). Because the enzymes catalyzing these reactions are participants in a normal metabolism, they usually only attack halogens at the particular positions on their substrates that happen to be made labile by the enzyme's normal action. Also, since the dehalogenation is a fortuitous feature of the reaction, there probably exist in normal metabolic pathways many more enzymes capable of such adventitious dehalogenation, but which have never been examined for this capacity, and for which a specific halogenated substrate has not yet arisen that would call attention to the adventitious reaction. Goldman (1972) has reviewed dehalogenation and listed several adventitious enzymatic dehalogenations.

Adventitious dehalogenations typically are found at early steps in a catabolic pathway. The deeper into a sequence of metabolic reactions, the less likely that an adventitious enzyme reaction will meet with a suitable halogenated substrate. This follows because a putative halogenated substrate for such an enzyme must enter a cell and be metabolized through each of the preceding enzymatic steps in the sequence to produce the compound acted on by the adventitious, dehalogenating enzyme. At each enzymatic step along such a path the halogen must be tolerated, from the initial permease onward. Frequently, it can be anticipated that the halogen substitution will cause the inhibition of one or more of these enzymatic steps, possibly even killing the cell by preventing the functioning of an essential enzyme. Thus, many of the adventitious dehalogenases so far recognized are found in the early steps of a catabolic pathway.

Recognizing this limitation points to an opportunity for the genetic engineer. Construction of an efficient pathway for the catabolism of a halogenated pollutant might involve the alteration of enzymatic steps preceding dehalogenation, rather than dehalogenation itself. These alterations would permit an organism's enzymes to tolerate and metabolize a halogenated pollutant until an intermediate was formed that could be dehalogenated by an adventitious or other dehalogenase.

New adventitious dehalogenation can also be created by genetic manipulation. One highly promising way to accomplish this is the site-directed mutagenesis of genes for enzymes catalyzing hydroxylation reactions. If mutant forms of such enzymes will tolerate a halogen at the position hydroxylated, the reaction product will be a chemically unstable halohydrin, which can decompose a halide and a ketone. Other enzymatic reactions, for instance hydrolysis and hydration reactions, may also be amenable to this approach.

Slater and Bull's (1982) observations that microbial consortia can arise in response to the strong selective pressures of a novel growth substrate might be rationalized if individual members of the consortium contribute essential, possibly adventitious, steps in a catabolic pathway, the product of such steps diffusing to the other members of the consortium. An excellent illustration of this is the degradation of 2-chloropropionamide by the consortium mentioned previously (Slater and Bull, 1982; Slater _et al._, 1979). The first step, a deamination yielding 2-chloropropionic acid, is carried out by a _Mycoplana_ sp. that gives up the product and derives no direct benefit from the reaction. Waste product(s) released from the species that degrade 2-chloropropionate sustain the _Mycoplana_ and some other ancillary species not essential to the degradation.

It is easy to imagine that introduction of a gene for the amidase into the chloropropionate degrading species would create a more efficient organisms, since diffusion of the 2-chloropropionate from _Mycoplana_ to the other community members would no longer be necessary. Moreover, such a recombinant organism would not lose its ability to mineralize the original substrate following the physical dispersal of the community. It should be possible for the genetic engineer to predict the success of interactions like those arising naturally in consortia and potentially to design and construct consortia and/or recombinant organisms performing novel and useful degradations.

An example of the prediction and construction of such a novel strain has been carried out by Knackmuss and his group (Reineke and Knackmuss 1978a, b, 1979; Hartmann _et al._, 1979; Knackmuss, 1981). These workers recognized a novel pseudomonad that could grow at the expense of 3- or 4-chlorocatechol, among other substrates, but which was unable to convert chlorobenzoates, except 3-chlorobenzoate, to these catechols. Simultaneously, they were aware that the benzoate dioxygenase specified by TOL plasmid would adventitiously accept several chlorobenzoate position isomers and convert them to chlorocatechols, including ones not attacked by the pseudomonad. By introducing the TOL plasmid into the pseudomonad and selecting strains growing at the expense of 4-chlorobenzoate, a novel strain was obtained. The strain not only expressed the appropriate enzyme activities specified by the TOL plasmid, but also apparently possessed regulatory mutations affecting both plasmid and chromosomal genes and causing the plasmid's benzoate oxygenase and the chromosome's catechol oxygenase and cycloisomerase enzymes to be expressed without interference from regulatory proteins controlling the related functions on plasmid and chromosome. They went on to select mutants extending the number of halobenzoates utilized by this strain. This was accomplished by feeding new substrates such as 3,5-dichlorobenzoate, along with a utilizable substrate, in a chemostat, and finding mutants that took over the culture because they could utilize both substrates.

The selective pressures leading to the isolation of mutants such as these, or to recombinant strains in associations like the chloropropionamide consortium, will ultimately depend upon a growth rate differential between the recombinant or mutant and the parental types. It is not always certain, however, that a single strain carrying an entire biodegradative pathway will grow more rapidly under a given set of conditions than will the members of a community, at least under the conditions that selected the community. Slater and Bull (1982) discuss a number of cases in which the community is apparently better adapted than a single recombinant strain.

The foregoing discussion is not meant to imply that the only fruitful approach to new biodegradative activities, such as new dehalogenation pathways, is through incisive, preplanned molecular manipulations. Useful, well adapted concatenations of biochemical steps still await isolation through traditional methods like enrichment and elective culture. For example, elective culture has recently yielded an *Arthrobacter* sp. that mineralizes pentachlorophenol using it as sole source of carbon and energy (Edgehill and Finn, 1982; Stanlake and Finn, 1982). Adaptations of enrichment culture for "plasmid-assisted breeding" have also been productive, as previously mentioned (Kellogg *et al*., 1981). These strains are of theoretical importance, offering new insights into enzymatic dehalogenations, as well as being of potential use in waste treatment.

Conclusions Concerning Natural Constraints of Biodegradation

A multitude of physical, chemical and biological factors can influence the rate and extent of biodegradation of environmental pollutants. While all of these should be considered in the design of an organism intended to address a particular pollutant in a particular habitat, this cannot currently be done. Not only is there too little information about the quantifiable characteristics of most polluted environments, there are factors such as potential interactions among resident and introduced microorganisms and the availability of suitable attachment sites for introduced strains that are unknown and essentially unknowable.

Given that a detailed analysis of the environmental factors limiting a biodegradative process is not currently feasible, it is nevertheless possible to prioritize the currently recognized constraints and to qualitatively assess their impact on the genetic engineering approach to pollution control.

Any attempt to do this must first recognize the object of creating a population of organisms of sufficient size and activity to effectively catalyze degradation of the target pollutant. The means proposed to accomplish this include (1) the introduction of a suitably engineered microbe into the polluted site, with the expectation that it will proliferate, and (2) the inoculation of the site with an organism carrying genes for a specific degradation on a transmissible plasmid that can pass into and acclimate the indigenous microflora. A third means, discussed in the next section, involves non-proliferating organisms. Thus the most important constraints to be considered are those controlling the proliferation of introduced microbes and their genes, and the accessibility of pollutants to microbes.

In soils the most important constraint on biodegradation is the inaccessibility of the pollutant to the microbe. Introduced microbes appear to be almost unable to move within a soil due to adsorption to soil particles and other factors. The absence of extensive movement restricts the pollutants that may be attacked by these microbes. Usually only the pollutants in the superficial soil horizon will be accessible.

Further study is needed to test the generality of this conclusion, and to explore means to overcome this barrier. Depending upon the outcome of such studies, it may be necessary to abandon the inoculation of soil as a proposed means to improve biodegradation, or at least to restrict it to particular strategies that overcome the mobility barrier. One such means might utilize root-colonizing bacteria in conjunction with the culture of an appropriate host plant at the polluted site to physically distribute the bacteria in the soil.

In sediments, landfills and wastewater treatment plants, the primary constraints seem to be those affecting a microbe's ability to survive, to proliferate and to function in a habitat. The constraints controlling a specific site can be any of those identified in this section. It is not possible in advance to identify which factor will be most important at any specific site, but it can be assumed that most factors will be essentially unalterable, except perhaps for the factors governing habitats in wastewater treatment systems. Thus, the problem reduces to developing a bacterium that can tolerate conditions in a particular habitat and that can also perform the desired transformation of the pollutant. In practice, finding a usable organism will probably involve testing manipulatable strains for survival in the habitat. Sometimes it may be possible to isolate a suitable species indigenous to a site and to attempt to manipulate it.

The choice between these two approaches will depend in large part on the complexity of the manipulation needed to endow an organism with the desired biodegradative capacity. This will depend upon a myriad of factors including the concentration of the target pollutant, the availability of the genes for the needed enzyme activities, details of gene transmission and gene expression in each potential host organism, and the potential for integration of the products of the new metabolism into a host's normal metabolism.

The strategy of spreading a relevant biodegradative capacity into an indigenous microflora by plasmid transmission is a plausible but untested idea. Since microbes do not move readily through soil, it would be novel indeed if plasmids were able to spread more readily than an introduced microbe. But this is nevertheless possible if conjugation can occur between interconnected, immobile indigenous bacterial colonies attached to the surfaces of soil particles. At present no attempts to evaluate this strategy have been undertaken. The influences of physical and biological factors on plasmid spread in the environment are completely unknown.

In summary, the consideration of natural factors limiting biodegradation in the environment suggests that the use of an organism tolerant of the prevailing physical and biological factors is a critical aspect of any new biological pollution control. Organisms unable to survive in the polluted habitat will never have the chance to express their biodegradative capacity, no matter how sophisticated the measures used to develop it.

5

Release to the Environment of Engineered Microorganisms

Introduction

The most important issue to emerge from this study concerns the release of altered microorganisms to the environment. Contact of microorganisms with environmental contaminants is inherent in virtually all deliberate biological waste treatment processes. Most biological treatment processes require contact of the waste with very large quanitites of metabolically active microbes in open systems. Treatment of a contaminated soil or sediment may actually require the establishment of an engineered strain in the environment, but even if establishment of a new microbe in the environment is not required for treatment, the use of an engineered strain means the release of the microbe to the environment and raises the possibility of its establishment there.

It is impossible to completely assure the containment of microbes and their DNA in full scale waste treatment systems. Even if consideration were restricted to a relatively contained system such as a wastewater treatment plant, where disinfection of sludge and effluents could be undertaken, it would be unrealistic to guarantee that no viable organisms would be released. For instance, Cronholm (1980) has detected wastewater organisms as much as 930 m downwind from treatment plants. Pereira and Benjaminson (1975) found wastewater organisms in air at a density of $17/m^3$ at a distance of 300 m from the aeration stacks of wastewater plants and reported that ozonation was a poor disinfectant in this case. Unlike the biological component of the wastewaters from pharmaceutical manufacturers, which may be contained and which is a reasonable model for the wastes from the emerging recombinant DNA industry, the sheer volume of effluent from typical municipal or industrial wastewater treatment facilities makes sterilization impossible. Economic constraints alone preclude this.

Moreover, basic studies indicate that simply killing microorganisms does not remove the possibility of transmission of the genetic information that they contain. For instance, under laboratory conditions certain wild type strains of the genus *Acinetobacter* have been shown to possess an extraordinary ability to be transformed by crude whole cell lysates of bacteria from the same genus (Sawula and Crawford, 1972; Juni, 1972). Certain strains of cyanobacteria are also readily transformed. Transformation ability is shared, to a lesser extent, by many Gram positive bacteria.

Therefore, the possibility of transmission of chemically intact DNA exists, even from dead organisms, although it remains to be shown that such transformation can occur in environmental matrices. It is imperative that any use of engineered organisms for waste treatment be considered a deliberate, irreversible release of the organism and its DNA to the environment.

The release of organisms containing DNA manipulated by the *in vitro* recombinant DNA techniques is already regulated under the NIH Guidelines for Research Involving Recombinant DNA Molecules (Department of Health and Human Services, 1982). Section III-A of these guidelines states that "Deliberate release into the environment of any organisms containing recombinant DNA ... cannot be initiated without submission of relevant information to NIH, the publication of the proposal in the *Federal Register* for thirty days of comment, review by the RAC (the Recombinant DNA Advisory Committee) and specific approval of NIH." These guidelines, although they do not have the force of law, provide a mechanism for the review and regulation of any release of microbes containing recombinant DNA created *in vitro*. On the other hand, they illustrate the seriousness attached to proposals involving the release engineered microorganisms to the environment.

These guidelines do not consider the release of microbes manipulated by *in vivo* recombination or by techniques such as mutagenesis. Some microbiologists view this as a serious oversight, especially regarding organisms that might be deliberately released to the environment in large numbers for pollution control. The USEPA apparently agrees, having recently established a committee to review the environmental safety aspects of research proposals involving genetically manipulated organisms.

The elements for a debate on environmental hazards accompanying the release of altered genotypes seem to lie within the fundamental viewpoints of microbial ecologists and molecular biologists. Microbial ecologists frequently observe and analyze changes in natural microbial populations in response to human interventions, including the large scale application of microbe-laden wastes to soils and sediments; molecular biologists frequently hold that microbial gene exchange is so rampant in nature that any combination of genes that can arise through normal physiological processes has already arisen during the history of the earth, and hence microbes carrying virtually all possible gene combinations have already been tested for the ability to establish themselves in the environment.

Both of these viewpoints are unconvincing in some situations. Microbial ecologists have produced a great deal of descriptive information on the environmental establishment of introduced microorganisms, but little of this information demonstrates the successful *prediction* of the establishment of any given microbe in any given habitat, as described in the next subsection. Admittedly, this is a very difficult task because of the many interdependent physical and biological factors that have an impact on an organisms's ability to become established in the environment. Thus, the potential for the establishment of novel organisms introduced into the environment is at present a matter of speculation. Molecular biologists, on the other hand, have detailed knowledge of gene exchange in only a very limited range of microbes, the best understood of which play very small roles in the environmental flora. For instance, the existence, let alone the frequency, of gene exchange among major microbial groups such as the eubacteria and the archaebacteria is currently unknown, yet the archaebacteria are ubiquitous in the environment and are responsible for the normal functioning of part of the carbon cycle on earth. The capacity for gene exchange among many autotrophic eubacteria as well as among many slow growing oligotrophic bacteria is also poorly documented. Thus, assertions that rampant gene exchange obviates risks from novel gene combinations are not actually as well documented as is frequently believed.

A plausible example will serve to illustrate the need for closer scrutiny. In Section 2, improvements in ammonia oxidation were mentioned as producing possible benefits in municipal wastewater treatment systems. The transfer of the nitrification functions from nitrifying autotrophs to heterotrophs residing in activated sludge aeration basins was one means suggested to accomplish the desired improvement. In reality, the number, location and organization of the genes responsible for nitrification, as well as the potential for normal gene transfer from species to species, is unknown in typical autotrophic nitrifiers such as *Nitrosomonas*. One could imagine that the low numbers of such organisms per unit volume in the environment, or restrictions imposed by their habitat, or the number and complexity of the essential nitrification genes, or the structure of their cell envelope, among many other unknown factors, effectively prevent the exchange of nitrification genes with heterotrophs under normal conditions. Yet, it is also easy to imagine that in high density pure culture, a broad host range, transfer derepressed (promiscuous) plasmid could be introduced into a laboratory strain of *Nitrosomonas*, and a derivative plasmid eventually isolated bearing all or some of the nitrification genes without resorting to *in vitro* recombinant DNA technology. A heterotrophic strain acquiring these genes might then be cultivated and deliberately released to an aeration basin to evaluate plasmid spread and its influence on nitrification. None of these manipulations would fall under the scrutiny of the RAC, according to its current mandate.

Yet, if not reviewed for environmental acceptability, this scenario, and a great many variations on it, might lead to serious, detrimental effects on the environment. It happens that normal autotrophic nitrifiers lose a small fraction of the ammonia that they oxidize and contribute it to the atmosphere as N_2O. Some microbiologists think that this contribution accounts for a major proportion of atmospheric N_2O (McElroy *et al.*, 1976; Bremner and Blackmer, 1981) although other scientists point to combustion sources (Weiss and Craig, 1976; Weiss, 1981). N_2O is responsible for a major fraction of the depletion of ozone in the stratosphere. In fact, the National Research Council (1982) has recently estimated that a doubling of N_2O would cause a 15 percent reduction in ozone. Further, it is well known that normal autotrophic nitrifiers greatly increase their production of N_2O when cultured in high density in the laboratory, and under certain other conditions (discussed in Bremner and Blackmer, 1981). It can be anticipated that changes in the intracellular environment of nitrification enzymes that would accompany the wide dissemination of nitrification genes by the conjugative transfer of a promiscuous plasmid to indigenous microorganisms, might mimic the conditions that increase the proportion of ammonia converted to N_2O by normal nitrifiers under stress in laboratory culture. The establishment of such heterotrophic N_2O producers might discernibly increase N_2O production on earth and hence influence the atmospheric ozone level.

Clearly this picture is laden with speculation, only some of which can be tested. Yet, the speculations are not totally implausible. How should the USEPA committee proceed in this and similar cases? The development of prototype organisms seems to be a minimum requirement for any meaningful evaluation of the risks attendant to the use of truly novel organisms. Most of the basic questions concerning the potential for gene exchange in the environment and the response of a strain to quantified environmental stresses can only be answered if the organism in question is available to test. Most questions of safety, performance and establishment in the environment can be safely addressed in limited, contained, pilot scale trials. It may be, however, that the development of strains perceived as potentially hazardous should take place under physical containment higher than the nominal P-1 levels currently required by the RAC for most work with non-pathogens, especially since the organisms used for the development of new pollution control techniques will probably be chosen for their high potential to survive in the environment. The assignment of an appropriate containment level could be the responsibility of a body like the EPA's newly formed committee, in consultation with the RAC. Whoever takes this responsibility, they should primarily consider the possibilities of detrimental effects and the potential for survival and establishment in the environment when assigning a containment level. For the reasons discussed in the subsection entitled "Recommendations Concerning the Fate of Released Organisms," the reviewing body should not attempt to weigh potential benefit against potential risk.

Background: The Fate of Released Microbes and Their Genes

Recent considerations of the fate of released microorganisms in the environment have focused on deleterious outcomes accompanying the escape of novel microbes containing recombinant DNA. Yet, as illustrated in Figure 4, this point of view contrasts sharply to the deliberate release of a microbe for the purposes of pollution control. The fate of a microbial population released to the environment will be determined by its ability to grow and divide under the prevalent conditions. Three behaviors can be readily recognized. In the first, the microbes are unable to reproduce, although they may be metabolically active. Such organisms may also be able to donate their DNA to other organisms by conjugation or transduction. Organisms in this state, however, would be unable to establish themselves in the environment, and therefore they might be viewed as particularly safe agents for pollution control. For instance, organisms that constitutively synthesize large amounts of a few enzymes useful for pollution abatement could be used as detoxification agents, as suggested in Section 2.

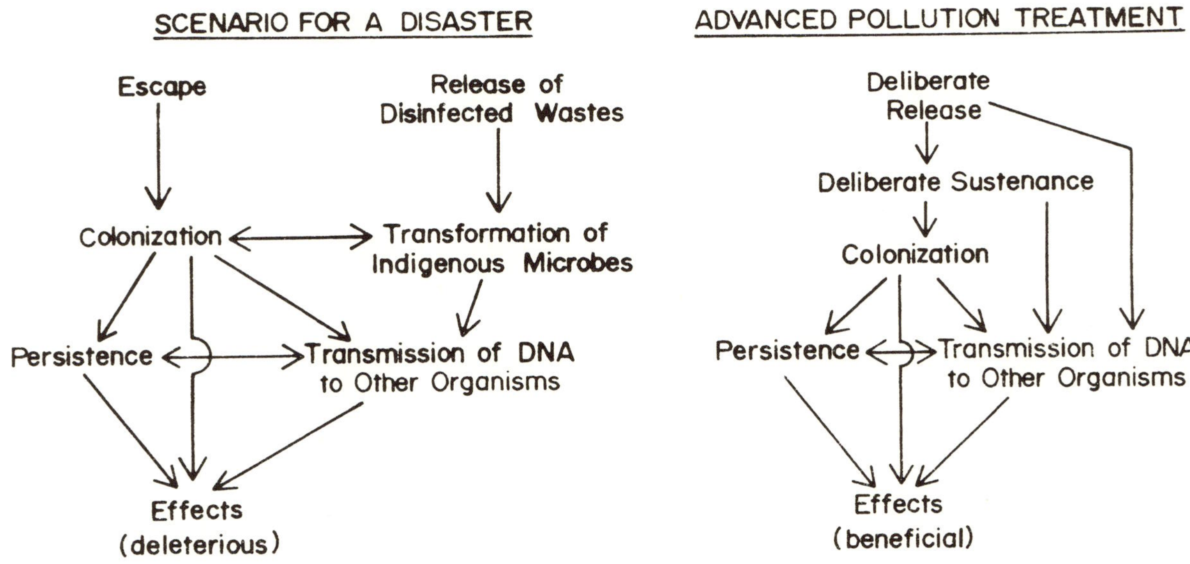

Figure 4. A comparison of terms used to describe the escape of a novel organism from a recombinant DNA laboratory and the use of a novel organism for pollution treatment.

The second discernible behavior of a microbial population includes the ability to reproduce, but an inability to establish a stable population. Such organisms also offer a degree of safety from unexpected environmental impacts, since they will not last indefinitely in the environment. Clearly, however, such organisms will have a much greater probability of transmitting their DNA to indigenous organisms and therefore may lead to the persistence of a genetic potential in the environment even if the original organisms themselves eventually die out. From studies of *Escherichia coli* (Marshall *et al.*, 1981; Levy *et al.*, 1980) described below, it seems likely that most microorganisms manipulated in the laboratory will exhibit one of these two behaviors.

The third discernible behavior is the ability both to reproduce and to persist essentially indefinitely in the environment. While there are only a few clear examples of introduced microorganisms persisting for a long time in the environment, for instance *Rhizobium japonicum* in soybean fields, there is as yet no reliable way to predict the potential for establishment of an introduced, manipulated strain.

There is a large body of literature concerned with the survival of microbes in various environments; it is typified by compendia such as *Survival of Vegetative Microbes* (Gray and Postgate, Eds., 1976) and reviews such as *Activity, Ecology and Population Dynamics of Microorganisms in Soil* (Stotzky, 1972), which focus on the degree of stress that particular organisms can withstand in particular environments. Studies such as these are beginning to identify the mechanisms of microbial inhibition and death in the environment. This information will be useful as it becomes possible to characterize the essential features of habitats and microbes that relate to polluted environments.

In the meantime, it is possible to glean some information concerning factors that regulate microbial proliferation in particular habitats from ecological studies, such as those reviewed by Alexander (1971; 1981). Coordinated studies attempting to characterize the influence of pH (Lowendorf, *et al.*, 1981; Mendez-Castro and Alexander, 1976), predation (Danso, *et al.*, 1975; Habte and Alexander, 1977; Ramirez and Alexander, 1980), moisture (Osa-Afiana and Alexander, 1979), the rhizosphere (Ramirez and Alexander, 1980) and transport (Madsen and Alexander, 1982) on *Rhizobium* illustrate the enormous effort needed just to begin to understand the factors governing a particular, economically important microbial population in soil. Even then, different species of *Rhizobium* show complex interactions with soil and rhizosphere that do not yet permit the prediction of success of a strain in new environments (Lowendorf *et al.*, 1981).

Much of the most recent information bearing on the fates of released organisms and their DNA comes from research primarily concerned with the containment of hosts and vectors for recombinant DNA. As a result, these studies deal mainly with debilitated enteric bacteria and their potential to colonize warm-blooded animals or wastewater sludges. Although this perspective is not wholly appropriate to waste treatment, useful information can still be gained, especially as it relates to the possibility of gene transfer.

An example is the study (Levy and Marshall, 1979; Levy et al., 1980) which tested the survival of E. coli x1776 and its pBR322-containing derivative, x2236, in the human intestinal tract. Neither strain colonized the intestine but the plasmid-containing strain persisted for a noticeably longer period than did its plasmid-less parent. This finding was unexpected since it is usually held that plasmid-bearing bacteria are at a competitive disadvantage relative to their plasmid-free derivatives (Helling et al., 1981; Wouters et al., 1980), at least in the absence of a selection pressure favoring plasmid-borne determinants. The authors suggested that unrecognized plasmid-encoded traits were responsible for the survival of x2236.

Another question raised was whether pBR322 was mobilized from x2236 during passage through the volunteer's intestinal tract. Fecal isolates phenotypically identified as x2236 were shown to be non-conjugative and to harbor normal pBR322. Other isolates exhibiting simultaneous resistance to ampicillin and tetracycline failed to hybridize to probes derived from pBR322. Thus, no mobilization of the plasmid was detected.

A related study characterized changes in the population of indigenous E. coli strains and in their plasmids in the human intestine over an eleven-month period (Caugant et al., 1981). By electrophoretic separation of the isozymes of 15 different enzymes, each of about 20 fecal isolates were characterized at every sampling. Fifty-three separate electrophoretic patterns (strains) were discerned during the study. The plasmid content of each electrophoretic type was also determined. Over the course of the study a few electrophoretic types predominated and were termed resident strains. Transient types were also present, persisting in the intestine for only a week to a month. The plasmid composition of the transient and resident populations changed frequently and radically. The authors explained the diversity of the flora over time by the invasion of new strain types. They felt in addition that the plasmid population was being strongly influenced by selection(s) for plasmid-encoded determinants, and that it did not merely reflect a simple balance between plasmid transmission and loss. This finding agreed with Levy and Marshall's (1979) conclusion that plasmids markedly affect the reproductive fitness of a strain. The study further suggested that "recombination occurs at a low rate and that populations exist and evolve as groups of nearly independent clones." Such a conclusion is surprising in view of earlier work by Levin and Stewart (1980) and Levin and Rice (1980) that described mathematical models for the kinetics of transfer and maintenance of non-conjugative plasmids in bacterial subpopulations even when they do not confer a selective advantage on their host.

While these studies refer to non-conjugative plasmids, organisms designed for pollutant transformations would not necessarily be similarly limited. Mathematical models of the kinetics of conjugative plasmid transfer have been published (Stewart and Levin, 1977; Levin et al., 1979). In addition, Sagoo and Cain (1978) have demonstrated empirically the conjugative transfer of a degradative plasmid under simulated environmental conditions. They examined transmission of ALS plasmid, carrying determinants for the metabolism of alkylbenzene sulfonates, in media extracted from sludge, soil, and marsh water, river and estuarine sea water. Good conjugation frequencies were found in all media but the latter two. Temperature and pH were found to be important variables in each case.

Studies of the survival of E. coli in wastewater sludge (Sagik and Sorber, 1979) and in lake sediments (LaLiberte and Grimes, 1982) show that this organism survives and divides in both media, although it does not produce a stable population. Such behavior offers sufficient opportunity for plasmid transfer. The transmission of antibiotic-resistance plasmids among enteric bacteria has been demonstrated in diffusion chambers suspended in various parts of wastewater treatment plants. Mach and Grimes (1982) demonstrated transmission in primary and the secondary clarifiers while Altherr and Kasweck (1982) showed transfer in influent wastewater but not in the treatment plant's effluent-receiving waters. In both studies, the donor strains and their plasmids were isolated from the wastewater treatment plants under consideration. In both studies, transfer in the laboratory gave at least ten-fold more recombinants than it did in the suspended diffusion chambers.

These are the first studies demonstrating transfer of plasmids among bacteria under conditions representative of normally operating wastewater treatment plants. They give hope of using plasmid spread to acclimate an indigenous microflora, but further study is needed.

Plasmid transmission has been directly confirmed in the human intestine, and it was found to occur much less frequently than anticipated from studies in pure culture (Marshall et al., 1981). Three plasmids, pBR322, pSL222-4 and pLM2, respectively non-conjugative, transfer-derepressed and broad host range (inc P), were administered to human volunteers. Strains expressing plasmid-determined antibiotic-resistance phenotypes were isolated from stool specimens and the isolates were examined for the presence of these plasmids. In particular, the study attempted to discern whether the conjugative plasmids could mobilize the non-conjugative cloning vector pBR322. To encourage the growth of plasmid-bearing strains as well as to encourage plasmid transfer, tetracycline was administered to the volunteers during part of the study. Neither the non-conjugative nor the broad host range plasmid was transferred at detectable frequencies. The transfer-derepressed plasmid was mobilized at frequencies about one one-hundred-thousandth that exhibited in laboratory culture; its transfer was enhanced by antibiotic administration. Also of note was the observation that the non-debilitated E. coli K-12 strain, hosting the plasmids, survived very poorly in the intestine, even when antibiotic was administered.

An integrated view of these studies suggests that plasmids transfer less readily in environmental matrices than under ideal laboratory conditions. Most of these studies have focused on typical enteric bacteria and antibiotic-resistance plasmids. Future work might expand consideration to other major species found in wastewater treatment sludges, biofilm reactors and anaerobic filters.

Research Recommendations Concerning the Fate of Released Organisms

Because the release of organisms is inherent in pollution control, the environmental fate of such microbes and their genes must be understood. There is a need for experimental study of the establishment of manipulated microbes and their genes introduced into polluted environments. There also is a need for a mechanism to review proposed releases and a policy to regulate them.

New research should be instituted to identify the barriers to colonization and gene transmission by introduced, engineered organisms in soils, sludges and sediments. Strategies to enhance plasmid transfer and/or colonization should be studied using bacterial species typical of the indigenous flora, not just emphasizing debilitated or novel enteric strains. At the same time, the efficiency of compound degradation by amendment of a polluted environment with specialized microbes that are not able to persist, should also be studied. Regarding the development of strains with specialized degradative functions, the present inadequate knowledge of factors governing a strain's capacity to become established suggest that minor modifications of strains taken from the target environment might offer the best means to obtain a new strain that can be reestablished in its original environment. In addition, the *in vitro* recombinant DNA methods should be used only when essential; *in vivo* techniques should be used whenever these can produce an equivalent result, even if the manipulation is somewhat more tedious. These two recommendations, the modification of indigenous strains and the emphasis on *in vivo* methods, may raise the probability of developing strains acceptable for use in the near future. Presumably as regulatory experience is gained and the biological factors determining the colonization of a habitat by an introduced strain are identified, it will become possible to routinely utilize the full range of available strains and techniques, including *in vitro* gene recombinations.

At present, it is desirable that a central regulatory committee review and sanction the experimental release of engineered pollution control microorganisms. This will permit scientific and policy judgements to be developed and experiences to be coordinated. The newly formed USEPA committee could well take on this function.

The development of organisms catalyzing novel biodegradations, and the experimental release of such organisms, will certainly be influenced by the USEPA position concerning actions involving real or predicted deleterious consequences. This position requires that potential risks must be weighed against estimates of benefit. Only situations that promise high potential benefit versus low potential risk are supposed to be supported.

The risk-benefit approach to decisions about research support is inappropriate to problems of pollution control using genetically manipulated organisms. This is because the perception of risks in this case depends on unreliable predictions of two extreme factors: the frequency of occurrence of very rare events and the severity of hypothetical environmental disasters. It is likely that the bulk of cases of deliberate release of novel microbes will always be susceptible to speculations about risks of this sort, leading to the argument that release should never be permitted.

It must be remembered that no action is ever completely free of risk. It is interesting to recall that a similar controversy arose over the improbable though plausible disaster predicted to accompany widespread vaccination of people with attenuated virus. Even though the benefits of vaccination are now clearly evident some controversy still continues regarding possible long term deleterious consequences of such vaccinations, for instance, in the case of polio vaccines potentially contaminated with transforming virus. It should be the responsibility of an oversight committee to wrestle with the probabilities, potentialities and policies relating to this issue.

It is recommended that the estimation of the scientific merit and the potential benefit of the release of an organism be performed separately from the estimation of the environmental risk. These evaluations should be performed by separate bodies acting in sequence. Potential benefit should be examined along with scientific merit, in confidence, as part of normal peer review; potential risk and assignment of containment level for approved proposals should be performed by a public body such as the newly formed USEPA committee. This separation is desirable because the normal balancing of benefit and risk is untenable for actions in which extremely rare but potentially disasterous events may occur.

Finally, it seems appropriate to remind scientists that permission to produce and distribute an extraordinary microbe is not a license to eschew ordinary professional practice. The culture and release of large quantities of any bacterial species must always be guided by practical but effective precautions that take into account all the known properties of the strain, not just its biodegradative capacity. Microbes to be manipulated genetically should be chosen and handled in such a way that they represent no threat to the health of those working with them or those that might subsequently contact them in the environment. In particular, all of the pathogenic properties of a species need to be considered.

6

Conclusions and Recommendations Concerning the Development of New Biological Pollution Control Agents

There was general agreement among the participants at the Urbana workshop regarding some aspects of the development of new biological pollution control agents. It was generally held that this approach offers a great deal of promise, although perhaps not as much as has been suggested in the sensationalist press. Several cautionary notes were sounded, however.

An obvious caution was that pollution problems must be clearly defined in terms relevant to genetics and biodegradation before potential solutions can be discussed. Seemingly self-evident, this notion is violated by suggestions such as the creation of "superbugs" to reduce the BOD of an effluent or the volume of sludge from an aeration basin. Genetic engineering most often involves the specific alteration of individual biochemical pathways in particular organisms. Individual strains with specialized pathways can have but little effect on the uncharacterized and presumably heterogeneous organic matter measured by the BOD test or present in wastewater sludge.

In a similar vein, a caution was expressed about the prospects for developing single organisms able to attack pollutants such as PCBs or toxaphenes, that are actually collections of diverse substances. It was pointed out that bacteria frequently attack only specific members of a set of related compounds. Thus, differences as small as a single hydroxyl substitution, for example in benzoic, *p*-hydroxybenzoic and salicylic acids, are discerned by bacteria. Usually such compounds are degraded by separate strains using wholly separate pathways. A single bacterial strain will frequently degrade compounds related as intermediates in a catabolic pathway, for example naphthalene, salicylate and catechol, more readily than a set of compounds with close chemical relationships, for instance, the position isomers phthalic and terephthalic acid, or a set of compounds with varying chemical substitutions at one position, such as chlorobenzene, phenol, aniline, nitrobenzene and toluene. This apparently stems from the generally high specificity of enzyme-substrate interactions and from the need to foster efficient degradation by integrating the products of catabolism into the central metabolism of an organism.

Apparently exceptions to this specificity do exist. Some oxygenases occurring early in a degradative sequence show a more catholic appetite than is usual for typical enzymes of intermediary metabolism, for example benzoate oxygenases that can accept chlorobenzoates as substrates (Reineke and Knackmuss, 1979) or naphthalene oxygenases that can utilize methyl naphthalenes. Further, it has been pointed out that many catabolic pathways tend to converge on a few common metabolites, for example, aromatic degradations converging on catechol or gentisate. Nevertheless, these exceptions to absolute specificity are not a sufficient basis to design an organism to attack so broad a categorical set of compounds as the PCBs; it must be anticipated that a large group of specifically developed bacterial strains, each attacking one or only a very few members of the set, will be needed to achieve a comprehensive attack on collections of compounds like these.

Another caution noted that compounds rarely occur singly in nature. Similarly, pollutants will most commonly be found as components of a mixture of compounds. The other compounds occurring in conjunction with a target pollutant might be either beneficial or detrimental to the organism degrading the pollutant. In addition, the physical conditions present at the polluted site will individually and perhaps synergistically alter the functioning of any given engineered microbe. Thus, a strain used at one site would not necessarily work well at another, even if both sites contain a common target compound. This would be due to the nature of the chemical mixture or the environmental conditions present at each site. Thus, at least initially, each pollution problem should be addressed on a site-by-site, pollutant-by-pollutant basis. Ultimately, it may be possible, after achieving improvements in treatments of some specific pollutants, to identify groups of sites with sufficiently similar characteristics that a particular manipulated organism can be used at all similar sites. The initial experimental work, however, should be done with a narrow focus to provide a foundation for what will follow.

Recommended research. At the present time, the strengthening of research in a few specific subjects appears to offer the most expeditious means to promote the genetic engineering approach to pollution control. The recommended research is tabulated in Table 7. It varies in scope from specific projects to broad fields of study. The table is organized by the breadth of definition of the research from broadest (fields and areas) to most narrowly defined (projects).

The establishment of introduced organisms in the environment and the transmission of genetic material among indigenous microflora are the two steps basic to every suggested improvement in pollution control through genetic engineering. The current state of understanding of these processes does not permit prediction of the success of either and does not provide the molecular biologist with guides for the construction of successful organisms. Study in these two fields was assigned the highest priority.

Many factors influence the potential for successful establishment of introduced organisms. These include physical environmental factors (pH, temperature, water activity, oxygen tension), nutritional factors and the presence of toxic or beneficial non-target compounds. Immobilization of the microbe at solid-liquid interfaces, adsorption and sequestration of target compounds at interfaces or by soluble polymers such as humic substances and lignins, and interactions of the introduced species with the indigenous microflora, especially competition and grazing by predatory microorganisms, must also play important roles.

Table 7
Outline of Conclusions and Recommendations

I. Release of Engineered Organisms for Pollution Control (Section 5)

A. Concern over the release of engineered organisms demands that experiments involving their release to the environment be reviewed and sanctioned by a scientifically qualified public body.

B. Current regulations governing the release of microbes altered by *in vitro* recombinant DNA methods are not a barrier to the development of such organisms, although release to the environment is regulated.

C. Development of new organisms by *in vivo* recombination and other techniques not now regulated should be emphasized.

D. The possibility of exceptional examples of potentially hazardous strains arising during the development of pollution control agents by purely unregulated manipulations cannot be ruled out at present.

II. General Conclusions (Section 6)

A. Genetic engineering offers promise for new pollution treatments.

B. Single strains degrading a categorical collection of pollutants (e.g., PCBs) will not be found.

C. Compound-specific and site-specific treatments must be emphasized initially.

III. Recommended Research

A. Fields
  i. Colonization of polluted environments by introduced microbes
  ii. Gene transmission in the environment
  iii. Co-metabolism
  iv. Oligotrophy

B. Areas
  i. The genetics of catabolism by archaebacteria
  ii. The genetics of catabolism by anaerobes
  iii. The genetics of catabolism by filamentous fungi
  iv. The genetics of microbial dehalogenation
  v. Artifical evolution of enzymes and metabolic pathways

C. Projects
  i. Flocculation in wastewater sludges
  ii. Ammonia removal by low-age wastewater sludge

Table 7 (continued)

IV. Implementation

A. Knowledge from relevant fields should be developed in parallel and integrated through a newsletter and problem-focused symposia.

B. Research undertaken by the USEPA and other government agencies and organizations should be coordinated.

C. A current listing of relevant scientific developments, pollution problems and research needs should be compiled, distributed and updated regularly.

It is not realistic to expect a refined knowledge of these factors to develop before any attempts are made to introduce organisms for environmental quality control. In the interim, the strategy of removing a stable species from a polluted site and altering it to enhance the destruction of a target compound is recommended as the most reasonable course. This manipulated strain should have a relatively high chance of being reintroduced into its site of origin and persisting, even if most of the factors responsible for its success are not known.

Gene transmission by plasmid spread, the other priority field of study, is already known to occur among bacteria in laboratory culture and in diffusion chambers suspended in wastewater treatment plants. However, this phenomenon has not been documented in open environmental systems with plasmids bearing catabolic pathways. It is known to occur much less readily in the human intestine than previously anticipated. Experiments to establish the existence of catabolic plasmid spread among microorganisms indigenous to wastewater sludges, anaerobic filters, soils, sediments, etc., should be sponsored. From such studies, the genetically controlled factors regulating plasmid spread may be identified and the technological aspects developed.

Establishment of an engineered strain and plasmid spread are both means to achieve an ehnanced degradation of a target compound. Each depends upon effective expression of certain genes either in the introduced strain or on the plasmid. To be effective, the genes must be compatible with the metabolic predisposition of the recipient strain and, in the case of plasmid-borne genes, these pathways should be compatible with the normal metabolism of a large number of recipient strains. Compatibility encompasses interactions of regulatory proteins providing for gene expression, functionality of gene products within the intracellular environment, and entrance of the products of the engineered catabolism into the central metabolism of the cell.

Although the need for such compatiblities is recognized, no specific recommendations to study problems of gene expression can be made, since these problems will arise on a case-by-case basis. A recommendation to study this topic in general is tantamount to recommending support for the entire area of the molecular biology of gene expression. Such a recommendation would have no specific impact on pollution control technology. The biochemists who reviewed the background papers believed that a good deal of the effort to develop new pollution control agents will be spent overcoming problems concerning gene expression and the regulation of metabolism. Any support for this work should be a part of the overall development of the strain of interest.

Other fields of research deserving support are catabolism by oligotrophic bacteria and co-metabolism. Co-metabolism offers significant promise for advanced waste treatment. A co-metabolic attack on a pollutant does not require that the target compound contribute to the growth or maintenance of the attacking microorganisms. Recently gained understanding of molecular genetics has suggested a variety of mechanisms by which co-metabolic phenomena may operate. It seems likely that many of these mechanisms might be manipulated genetically to adapt the phenomenon to the treatment of specific pollutants. In particular, the isolation of mutants constitutively producing a relevant catbolic pathway, in conjunction with site-specific mutagenesis to modify enzymes in the pathway so that they may accept target compounds related to their normal substrates, appear to offer substantial promise.

Oligotrophic bacteria are those that colonize habitats with persistently low nutrient flux. They survive in very low concentrations of nutrient compounds, producing an abundance of high affinity transporter proteins and a streamlined, highly efficient energy metabolism. Since a great many persistent pollutants exist in the environment at very low concentrations, it seems desirable to study and understand organisms specifically adapted to low nutrient environments, with the aim of eventually adapting their strategies to the treatment of dilute pollutants. Within the fields of co-metabolism and oligotrophy, there are a great many specific projects that can be envisioned that would advance knowledge of each of these fields.

More narrowly defined than the fields of research above, five areas of research are recommended for support: 1) the metabolism and genetics of archaebacteria, 2) of anaerobes, 3) of filamentous fungi, 4) of organisms catalyzing dehalogenation reactions, and 5) the artificial evolution of enzymes and metabolic pathways. Each of these areas is described in the previous sections. They are underexplored and offer important opportunities for the development of new strains having desirable catabolic activities. In each case, some preliminary or exploratory research has been done, but substantial voids in knowledge and experience prevent their routine use for strain development.

Specific projects recommended include the improvement of flocculation and settling of organisms in wastewater sludge and improvements in the oxidation of ammonia under heterotrophic conditions in aeration basins. The flocculation and settling of wastewater sludges is an important process limiting the quality of effluents. It involves the adhesion of microorganisms to themselves and to solid surfaces, which depends upon both the structure of the microbial cell envelope and the production of glycoproteins to condition solid surfaces. If these factors are under genetic control, then improvements through genetic intervention are possible, in principle. Exploration of the genetic factors influencing attachment may reveal means to improve fixed-film bioreactors or to physically join members of microbial consortia.

The enhancement of ammonia oxidation initially involves finding the factors that regulate the reproduction of autotrophic nitrifiers. The problem appears to be complex, involving the fundamental nature of autotrophy. Work in this area is also subject to speculations about deleterious environmental impacts, as described earlier. Even a very limited study of a few key properties of these autotrophs, for instance their inherent capacity to exchange DNA with environmental heterotrophs, might allay many of the fears suggested by this speculation, or reinforce them. Considering both the potential benefits and the potential risks, it seems highly desirable to undertake such a study.

It is apparent from what has preceded that a broad range of scientific disciplines must contribute to the development of any truly workable new treatment. The fields of biochemistry, environmental engineering, microbial genetics, microbial physiology, ecology and molecular biology must all be represented and contribute. It is also desirable that relevant contributions be integrated effectively. At the moment, information concerning microbial ecology and the physiology of microorganisms lags behind the state-of-the-art of gene manipulation. Specifically, basic research on the colonization of new habitats by introduced organisms and on gene transmission in the environment needs to be emphasized. The integrated development of new knowledge would be fostered by the facilitation of an information exchange between interested scientists. An occasional newsletter, perhaps modelled on the aquatic microbiology newsletter published by members of the American Society of Microbiology, and pollutant-focused symposia, are two vehicles that might be particularly useful. One workshop participant recommended that these symposia should select a single pollutant and bring all of the relevant information from each discipline to bear on a discussion of potential new methods of treatment. Symposium topics might possibly be restricted to a discussion of the pollutant in a single setting or at a particular site. The proposed newsletter could also be used to identify individuals in each relevant discipline with specific knowledge and interest in the interdisciplinary approach to the development of new pollution controls.

The newsletter could be a vehicle for communication among scientists and engineers involved in the improvement of pollution control. It might compile and annually revise a listing of relevant sponsored research. News of significant new developments could be reported, meetings could be advertised and a forum for comments or debate could be provided in its pages. If it were widely used, it could help to breach the existing barriers created by misconceptions, jargon and divergent disciplinary perspectives.

Concluding Remarks. Molecular biologists have created a set of powerful new tools for the manipulation of genetic information. It is up to the scientists and engineers who deal with the practical aspects of biology to discern the applications that will benefit man and the environment.

New pollution controls are likely to emerge from the interaction of basic biological research, environmental engineering and gene manipulation. At present, the new gene manipulation technologies are having little impact on practical problems of pollution control because essential basic information is missing concerning the nature of genetically controlled functions limiting biodegradation in the environment. As knowledge of these factors is refined, it can be anticipated that new and improved pollution controls will follow.

References

Alexander, M. Biochemical ecology of microorganisms. Ann. Rev. Microbiol., 25:361-392, 1971.

Alexander, M. Why microbial predators and parasites do not eliminate their prey and hosts. Ann. Rev. Microbiol., 35:113-133, 1981.

Allan, J. Loss of biological efficiency of cattle-dipping wash containing benzene hexachloride. Nature London, 175:1131-1132, 1955.

Altherr, M. R. and K. L. Kasweck. *In situ* studies with membrane diffusion chambers of antibiotic resistance transfer in *Escherichia coli*. Appl. Environ. Microbiol., 44:838-843, 1982.

Alvarado-Urbina, G., G. M. Sathe, W-C. Liu, M. F. Gillen, P. D. Duck, R. Bender and K. K. Ogilvie. Automated synthesis of gene fragments. Science, 214:270-274, 1981.

Anderson, J. J. and S. Dagley. Catabolism of aromatic acids in *Trichosporon cutaneum*. J. Bacteriol., 141:534-543, 1980.

Andrade, P. S. L. and D. A. Carlson. Identification of a Mirex metabolite. Bull. Environ. Contam. Toxicol., 14:473-479, 1975.

Atlas, R. M. Effects of temperature and crude oil composition on petroleum biodegradation. Appl. Microbiol., 30:396-403, 1975.

Atlas, R. M. Fate and Effects of Oil Pollutants in Extremely Cold Marine Environments. U.S. Dept. of Commerce National Technical Information Service. AD-A-070-992, 1979.

Bacila, M., B. L. Horecker and A. O. M. Stoppani. Biochemistry and Genetics of Yeasts, Pure and Applied Aspects. New York: Academic Press, 1978.

Baess, I. Report on a pseudolysogenic *Mycobacterium* and a review of the literature concerning pseudolysogeny. Acta Path. Microbiol. Scand., 79:428-434, 1971.

Bagdasarian, M. and K. N. Timmis. Host: Vector Systems for Gene Cloning in *Pseudomonas*. In: Current Topics in Microbiology and Immunology, Vol. 96, W. Henle, P. H. Hofschneider, H. Koprowski, F. Melchers, R. Rott, H. G. Schweiger and P. K. Vogt, eds. Heidelberg, Springer-Verlag, pp. 47-67, 1982.

Bagdasarian, M., R. Lurz, B. Ruckert, F. C. H. Franklin, M. M. Bagdasarian, J. Frey and K. N. Timmis. Specific-purpose plasmid cloning vectors II. Broad host range, high copy number, RSF 1010 - derived vectors, and a host-vector system for gene cloning in Pseudomonas. Gene, 16:237-247, 1981.

Baker, P. B. and D. R. Woods. Co-metabolism of the ixodicide Amitraz. J. Appl. Bacteriol., 42:187-196, 1977.

Baltz, R. H. Genetic recombination in Streptomyces fradiae by protoplast fusion and cell regeneration. J. Gen. Microbiol., 107:93-102, 1978.

Barder, P. S., F. O. Morrison and R. S. Whitaker. Conversion of DDT to DDD by Proteus vulgaris, a bacterium isolated from the intestinal flora of a mouse. Nature London, 205:621-622, 1965.

Barnsley, E. A. The bacterial degradation of fluoranthene and benzo(a)pyrene. Can. J. Microbiol., 21:1004-1008, 1975.

Bartha, R. Pesticide interaction creates hybrid residue. Science, 166:1299-1300, 1969.

Beam, H. W. and J. J. Perry. Co-metabolism as a factor in microbial degradation of cycloparaffinic hydrocarbons. Arch. Mikrobiol., 91:87-90, 1973.

Beam, H. W. and J. J. Perry. Microbial degradation of cycloparaffinic hydrocarbons via co-metabolism and commensalism. J. Gen. Microbiol., 82:163-169, 1974.

Bell, G. R. Some morphological and biochemical characteristics of a soil bacterium which decomposes 2,4-dichlorophenoxyacetic acid. Can. J. Microbiol., 3:821-840, 1957.

Bergmans, H. E. N., D. H. Kooijman and W. P. M. Hoekstra. Cotransformation of linear chromosomal DNA and plasmid DNA in Escherichia coli. FEMS Microbiol. Lett., 9:211-214, 1980.

Beringer, J. E., J. L. Beynon, A. V. Buchanan-Wollaston and A. W. B. Johnston. Transfer of the drug-resistance transposon Tn5 to Rhizobium. Nature, London, 276:633-634, 1978.

Berkeley, R. C. W., J. M. Lynch, J. Melling, P. R. Rutter and B. Vincent, eds. Microbial Adhesion to Surfaces. Ellis Horwood, Ltd., Chichester, 1980.

Bramucci, M. G. and P. S. Lovett. Selective plasmid transduction in B. pumilus. J. Bacteriol., 131:1029-1032, 1977.

Bibb, M. J., J. M. Ward and D. A. Hopwood. Transformation of plasmid DNA into Streptomyces at high frequency. Nature, 274:398-400, 1978.

Bird, J. A. and R. B. Cain. Microbial degradation of alkylbenzenesulphonates: Metabolism of homologues of short alkyl chain length by an Alcaligenes sp. Biochem. J., 140:121-134, 1974.

Birnboim, H. C. and J. Doly. A rapid alkaline extraction procedure for screening recombinant plasmid DNA. Nucl. Acids Res., 7:1513-1523, 1979.

Bollag, J. M., C. S. Helling and M. Alexander. Enzymatic hydroxylation of chlorinated phenols. J. Agr. Food Chem., 16:826-828, 1968.

Bollen, W. B. Interactions between pesticides and soil microorganisms. Ann. Rev. Microbiol., 15:69-92, 1961.

Bremner, J. M. and A. M. Blackmer. Terrestrial Nitrification as a Source of Atmospheric Nitrous Oxide. In: Denitrification, Nitrification and Atmospheric Nitrous Oxide. C. C. Delwiche, ed. John Wiley and Sons, New York, pp. 151-170, 1981.

Brock, T. D. Biology of Microorganisms, 3rd ed. Prentice-Hall Inc., New Jersey, pp. 625-626 and 777, 1979.

Brock, T. D., K. M. Brock, R. T. Belly and R. L. Weiss. Sulfolobus: a new genus of sulfur-oxidizing bacteria living at low pH and high temperature. Arch. Mikrobiol., 84:54-68, 1972.

Brown, A. D. Microbial water stress. Bacteriol. Rev., 40:803-846, 1976.

Brownell, G. H. and J. N. Adams. Linkage and segregation of a mating type specific phage and resistance characters in nocardial recombinants. Genetics, 60:437-448, 1968.

Bryant, M. P., E. A. Wolin, M. J. Wolin and R. S. Wolfe. Methanobacillus omelianskii a symbiotic association of two species of bacteria. Arch. Mikrobiol., 59:20-31, 1967.

Bukhari, A. I., J. A. Shapiro and S. L. Adhya. DNA Insertion Elements, Plasmids and Episomes, Cold Spring Harbor Lab., N.Y., pp 139-248, 1977.

Burns, R. G. Interaction of Microorganisms, Their Substrates and Their Products with Soil Surfaces. In: Adhesion of Microorganisms to Surface. D. C. Ellwood, J. Melling and P. Rutter, eds., Academic Press, New York, 109-138, 1979.

Burns, R. G. Microbial Adhesion to Soil Surfaces: Consequences for Growth and Enzyme Activities. In: Microbial Adhesion to Surfaces. R. C. W. Berkeley, J. M. Lynch, J. Melling, P. R. Rutter and B. Vincent, eds., Ellis Horwood, Ltd., Chichester, pp. 249-262, 1980.

Capet-Antonini, F. and E. Mankiewicz. Lytic phages produced by DNA-infected Mycobacteria. Can. J. Microbial., 14:647-652, 1968.

Cappenberg, T. E. A study of mixed continuous cultures of sulfate-reducing and methane-producing bacteria. Microbial Ecology, 2:60-72, 1975.

Carlberg, S. R. Oil Pollution of the Marine Environment -- with an Emphasis on Estuarine Studies. In: Chemistry and Biogeochemistry of Estuaries. E. Olausson and I. Cato, eds., Wiley & Sons, Ltd., New York, pp. 367-402, 1980.

Casida, L. E. and K.-C. Liu. Arthrobacter globiformis and its bacteriophage in soil. Appl. Microbiol., 28:951-959, 1974.

Castro, C. E. and N. O. Belser. Biodehalogenation. Reductive dehalogenation of biocides ethylene dibromide, 1,2-dibromo- 3-chloropropane and 2,3-dibromobutane in soil. Environ. Sci. Technol., 2:779-783, 1968.

Cattabeni, F., A. Cavallaro and G. Galli. Dioxin: Toxicological and Chemical Aspects. Halsted Press, New York, 1978.

Caugant, D. A., B. R. Levin and R. K. Selander. Genetic diversity and temporal variation in the E. coli population of a human host. Genetics 98:467-490, 1981.

Cerniglia, C. E. and D. T. Gibson. Metabolism of naphthalene by Cunninghamella elegans. Appl. Environ. Microbiol., 34;363-370, 1977.

Cerniglia, C. E. and D. T. Gibson. Oxidation of Benzo(a)pyrene by the filamentous fungus Cunninghamella elegans. J. Biol. Chem., 254: 12174-12180, 1979.

Cerniglia, C. E., D. T. Gibson and C. Van Baalen. Oxidation of naphthalene by cyanobacteria and microalgae. J. Gen. Microbiol., 116:495-500, 1980.

Chacko, C. I., J. L. Lockwood and M. Zabik. Chlorinated hydrocarbon pesticides: Degradation by microbes. Science, 154:893-895, 1966.

Chakrabarty, A. M. Transcriptional Control of the Expression of a Degradative Plasmid in Pseudomonas. In: Control of Transcription. B. B. Biswas, R. K. Mandal, A. Stevens, and W. E. Cohn, eds., Plenum, New York, pp. 157-165, 1974.

Chakrabarty, A. M. Plasmids in Pseudomonas. Ann Rev. Genet., 10:7-30, 1976.

Chakrabarty, A. M., G. Chou and I. C. Gunsalus. Genetic regulation of octane dissimilation plasmid in Psuedomonas. Proc. Natl. Acad. Sci. U.S., 70:1137-1140, 1973.

Chakrabarty, A. M., D. A. Friello and L. H. Bopp. Transposition of plasmid DNA segments specifying hydrocarbon degradation and their expression in various microorganisms. Proc. Natl. acad. Sci. U.S., 75:3109-3112, 1978.

Chakrabarty, A. M., C. F. Gunsalus and I. C. Gunsalus. Transduction and the clustering of genes in fluorescent pseudomonads. Proc. Natl. Acad. Sci. U.S., 60:168-175, 1968.

Chater, K. F., D. A. Hopwood, T. Kieser and C. J. Thompson. Gene Cloning in Streptomyces. In: Current Topics in Microbiology and Immunology, Vol. 96. W. Henle, P. H. Hofschneider, H. Koprowski, F. Melchers, R. Rott, H. G. Schweiger and P. K. Vogt, eds., Springer-Verlag, New York, pp. 69-95, 1982.

Chatterjee, D. K., J. J. Kilbane and A. M. Chakrabarty. Biodegradation of 2,4,5-trichlorophenoxyacetic acid in soil by a pure culture of Pseudomonas cepacia. Appl. Environ. Microbiol., 44:514-516, 1982.

Chou, J., P. G. Lemaux, M. J. Casadaban and S. N. Cohen. Transposition protein of Tn3: Identification and characterization of an essential repressor-controlled gene product. Nature London, 282:801-806, 1979.

Clark, D. L., A. A. Weiss, and S. Silver. Mercury and organomercurial resistances determined by plasmids in Pseudomonas. J. Bacteriol., 132:186-196, 1977.

Clarke, D. D., W. J. Nichlas and J. Palumbo. Fluorofumarate -- A substrate for fumarate hydratase. Arch. Biochem. Biophys., 123:205-206, 1968.

Clarke, P. H. The Evolution of Enzymes for the Utilisation of Novel Substrates. In: Evolution in the Microbial World. Soc. for Gen. Microbiol., Cambridge University Press, pp. 183-218, 1974.

Clowes, R. C. Molecular structure of bacterial plasmids. Bacteriol. Rev., 36:361-405, 1972.

Cohen, S. N., A. C. Y. Chang, and L. Hsu. Nonchromosomal antibiotic resistance in bacteria: Genetic transformation of Escherichia coli by R-factor DNA. Proc. Natl. Acad. Sci. U.S., 69:2110-2114, 1972.

Connors, M. A. and E. A. Barnsley. Naphthalene plasmids in pseudomonads. J. Bacteriol., 149:1096-1101, 1982.

Cornelis, G., D. Ghosal and H. Saedler. Tn951: A new transposon carrying a lactose operon. Molec. Gen. Genet., 160:215-224, 1978.

Cosgrove, D. J. Microbial Transformations in the Phosphorus Cycle. In: Advances in Microbial Ecology, Vol. 1, M. Alexander, ed., Plenum Press, New York, pp. 95-134, 1977.

Crawford, J. T. and J. H. Bates. Isolation of plasmids from mycobacteria. Infection and Immunity, 24:979-981, 1979.

Crockett, J. K. and G. H. Brownell. Isolation and characterization of a lysogenic strain of Nocardia erythropolis. J. Virol., 10:737-745, 1972.

Cronholm, L. S. Potential health hazards from microbial aerosols in densely populated urban regions. Appl. Environ. Microbiol., 39:6-12, 1980.

Crosa, J. H. and S. Falkow. Plasmids. In: Manual of Methods for General Bacteriology. Am. Soc. Microbiol., Washington, DC, pp. 266-284, 1981.

Dalton, H. and D. I. Stirling. Co-metabolism. Phil. Trans. R. Soc. Lond., B297:481-496, 1982.

Danso, S. K. A., S. O. Keya and M. Alexander. Protozoa and the decline of Rhizobium populations added to soil. Can. J. Microbiol., 21:884-895, 1975.

Dart, E. C., D. Pioli and K. T. Atherton. Genetic Engineering. In: Essays in Applied Microbiology. J. R. Norris and M. H. Richmond, eds., John Wiley and Sons, New York, pp 3/1-3/32, 1981.

Davis, E. M., J. Bishop and R. K. Guthrie. Resistance of Pollutants to Degradation in Saline Environments. In: Microbial Degradation of Pollutants in Marine Environments. A. W. Bourquin and P. H. Pritchard, ed., USEPA report 600/9-79-012, pp. 337-347, 1979.

Davis, J. B. and R. L. Raymond. Oxidation of alkyl-substituted cyclic hydrocarbons by a Norcardia during growth on m-alkanes. Appl. Microbiol., 9:383-388, 1961.

Davis, R. W., D. Botstein and J. R. Roth. Advanced Bacterial Genetics, A Manual for Genetic Engineering. Cold Spring Harbor Laboratory, Cold Spring Harbor, NY, pp. 116-125, 1980.

Dean, D. A plasmid cloning vector for the direct selection of strains carrying recombinant plasmids. Gene 15:99-102, 1981.

Department of Health and Human Services, National Institute of Health. Recombinant DNA research, proposed revised guidelines. Federal Register 43(146):33049-33050, 1978.

Department of Health and Human Services, National Institute of Health. Guidelines for research involving recombinant DNA molecules. Federal Register, 47(167):38049-38050, section III-A, 1982.

di Domenico, A., V. Silano, G. Viviano and G. Zapponi. Accidental release 2,3,7,8-Tetrachlorodibenzo-p-dioxin (TCDD) at Sevesco, Italy V environmental persistence of TCDD in soil. Ecotoxicol. Environ. Safety, 4:339-345, 1980.

Dimitriadis, G. J. Entrapment of plasmid DNA in liposomes. Nucl. Acids Res., 6:2697-2705, 1979.

Don, R. H. and J. M. Pemberton. Properties of six pesticide degradation plasmids isolated from Alcaligenes paradoxus and Alcaligenes eutrophus J. Bacteriol., 145:681-685, 1981.

Dow, C. S. and R. Whittenbury. Prokaryotic Form and Function. In: Contemporary Microbial Ecology, D. C. Ellwood, J. N. Hedges, M. J. Latham, J. M. Lynch and J. H. Slater, eds., Academic Press, NY, pp. 391-417, 1980.

Dubnau, D. Genetic Transformation of Bacillus subtilis: A Review with Emphasis on the Recombination Mechanism. In: Microbiology - 1976, D. Schlessinger, ed., Am. Soc. for Microbiology, Washington, DC, pp. 14-27, 1976.

Dunn, N. W., H. M. Dunn and R. A. Austen. Evidence for the existence of two catabolic plasmids coding for the degradation of naphthalene. J. Gen. Microbiol., 117:529-533, 1980.

Dunn, N. W. and I. C. Gunsalus. Transmissible plasmid coding early enzymes of naphthalene oxidation in Pseudomonas putida. J. Bacteriol., 114:974-979, 1973.

Edgehill, R. U. and R. K. Finn. Isolation, characterization and growth kinetics of bacteria metabolizing pentachlorophenol. European J. Appl. Microbiol. Biotechnol., 16:179-184, 1982.

Ehrlich, S. D. and A. Goze. Expression of Foreign Genes. In: Recombinant DNA and Genetic Experimentation, J. Morgan and W. J. Whelan, ed., Pergamon Press, New York, pp. 109-118, 1979.

Ehrlich, S. D. and V. Sgaramella. Barriers to the heterospecific gene expression among prokaryotes. Trends in Biochem. Sci. 3:259-261, 1978.

Ellwood, D. C., J. Melling and P. Rutter, Eds. Adhesion of Microorganisms to Surfaces. Academic Press, New York, 1979.

Engelhardt, G., H. G. Rast and P. R. Wallnofer. Bacterial metabolism of substituted phenols. Arch. Microbiol., 114:25-33, 1977.

Engesser, K.-H., E. Schmidt, and H.-J. Knackmuss. Adaptation of Alcaligenes eutrophus B9 and Pseudomonas sp. B13 to 2-Fluorobenzoate as growth substrate. Appl. Environ. Microbiol., 39:68-73, 1980.

Evans, W. C. Biochemistry of the bacterial catabolism of aromatic compounds in anaerobic environments. Nature, London, 270:17-22, 1977.

Ferry, J. G. and R. S. Wolfe. Anaerobic degradation of benzoate to methane by a microbial consortium. Arch. Microbiol., 107:33-40, 1976.

Finger, J. and V. Krishnapillai. Host range, entry exclusion and incompatibility of Pseudomonas aeruginosa FP plasmids. Plasmid, 3:332-342, 1980.

Fisher, P. R., J. Appleton, and J. M. Pemberton. Isolation and characterization of the pesticide-degrading plasmid pJP1 from Alcaligenes paradoxus. J. Bacteriol., 135:798-804, 1978.

Fisher, N. S., E. J. Carpenter, C. C. Remsen and C. F. Wurster. Effects of PCB on interspecific competition in natural and gnotobiotic phytoplankton communities in continuous and batch cultures. Microbiol. Ecology., 1:39-50, 1974.

Fletcher, M. The Attachment of Bacteria to Surfaces in Aquatic Environments. In: Adhesion of Microorganisms to Surfaces, D. C. Ellwood, J. Melling and P. Rutter, ed., Academic Press, New York, pp. 87-108, 1979.

Fletcher, M., M. J. Latham, J. M. Lynch and P. R. Rutter. The Characteristics of Interfaces and their Role in Microbial Attachment. In: Microbial Adhesion to Surfaces, R. C. W. Berkeley, J. M. Lynch, J. Melling, P. R. Rutter and B. Vincent, eds., Ellis Horwood, Ltd., Chichester, pp. 67-76, 1980.

Fliermans, C. B. and T. D. Brock. Ecology of sulfur-oxidizing bacteria in hot acid soils. J. Bacteriol., 111:343-350, 1972.

Focht, D. D. Mixed culture degradation of DDT metabolites. Bacteriol. Proc., pp. 16, 1971.

Focht, D. D. Microbial degradation of DDT metabolites to carbon dioxide, water and chloride. Bull. Environ. Contam. Toxicol., 2:52-56, 1972.

Fodor, K. and L. Alfoldi. Fusion of protoplasts of Bacillus megaterium. Proc. Natl. Acad. Sci. U.S., 73:2147-2150, 1976.

Fodor, K. and L. Alfoldi. Polyethylene-glycol induced fusion of bacterial protoplasts - Direct selection of recombinants. Molec. Gen. Genet., 168:55-59, 1979.

Foster, J. W. Bacterial Oxidation of Hydrocarbons. In: Oxygenases, O. Hayaishi, ed., Academic Press, New York, pp. 241-271, 1962.

Fox, G. E., E. Stackebrandt, R. B. Hespell, J. Gibson, D. Maniloff, T. A. Dyer, R. W. Wolfe, W. E. Balch, R. S. Tanner, L. J. Magnum, L. B. Zablen, R. Blakemore, R. Gupta, L. Bonen, B. J. Lewis, D. A. Stahl, K. R. Luehrsen, K. N. Chen and C. R. Woese. The phylogeny of prokaryotes. Science 209:457-463, 1980.

Francis, J. C. and P. E. Hansche. Directed evolution of metabolic pathways in microbial populations. I. modification of the acid phosphatase pH optimum in S. Cerevisiae. Genetics. 70:59-73, 1972.

Franklin, F., M. Bagdasarian and K. Timmis. Manipulation of Degradative Genes of Soil Bacteria. In: Microbial Degradation of Xenobiotics and Recalcitrant Compounds, T. Leisinger, R. Hutter, A. M. Cook, and J. Nuesch, eds., Academic Press, London, pp. 109-130, 1981.

French, A. L. and R. A. Hoppingarner. Dechlorination of DDT by membranes isolated from E. coli. J. Econ. Entomol., 63:756-759, 1970.

Friello, D. A., J. R. Mylroie, D. T. Gibson, J. E. Rogers and A. M. Chakrabarty. XYL, a nonconjugative xylene-degradative plasmid in Pseudomonas Pxy. J. Bacteriol., 127:1217-1224, 1976.

Fries, G. F., G. S. Marrow and C. H. Gordon. Metabolism of o,p'- and p,p'-DDT by rumen microorganisms. J. Agr. Food Chem., 17:860-862, 1969.

Fuhs, G. W. and M. Chen. Microbiological basis of phosphate removal in the activated sludge process for the treatment of wastewater. Microbial Ecology, 2:119-138, 1975.

Furukawa, K., F. Matsumura and K. Tonomura. Alcaligenes and Acinetobacter strains capable of degrading polychlorinated biphenyls. Agr. Biol. Chem., 42:543-548, 1978.

Furukawa, K., N. Tomizuka, and A. Kamibayashi. Effect of chlorine substitution on the bacterial metabolism of various polychlorinated biphenyls. Appl. Environ. Microbiol., 38:301-310, 1979.

Gabor, M. H. and R. D. Hotchkiss. Parameters governing bacterial regeneration and genetic recombination after fusion of Bacillus subtilis protoplasts. J. Bacteriol., 137:1346-1353, 1979.

Gaudy, A. F. and E. T. Gaudy. Microbiology for Environmental Scientists and Engineers. McGraw-Hill, New York, 1980.

Gautier, F. and R. Bonewald. The use of plasmid R1162 and derivatives for gene cloning in the methanol-utilizing Pseudomonas AM1. Molec. Gen. Genet., 178:375-380, 1980.

Gelbart, S. M. and S. E. Juhasz. Genetic transfer in Mycobacterium phlei. J. of Gen. Microbiol., 64:253-254, 1970.

Gelbart, S. M. and S. F. Juhasz. Transduction in Mycobacterium phlei. Antonie van Leeuwenhoek, 39:1-10, 1973.

Gill, R. E., F. Heffron and S. Falkow. Identification of the protein encoded by the transposable element Tn3 which is required for its transposition. Nature London 282:797-801, 1979.

Goldman, P. The enzymatic cleavage of the carbon-fluorine bond in fluoroacetate. J. Biol. Chem., 240:3434-3438, 1965.

Goldman, P. The carbon-fluorine bond in compounds of biological interest. Science, 164:1123-1130, 1969.

Goldman, P. Enzymology of Carbon-Halogen Bonds. In: Degradation of Synthetic Organic Molecules in the Biosphere, I. C. Gunsalus, ed., National Academy of Sciences, Washington, D.C., pp. 147-165, 1972.

Goldstein, A. The mechanism of enzyme-inhibitor-substrate reactions, illustrated by the chlorinesterase-physostigmine-acetylcholine system. J. Gen. Physiol., 27:529-580, 1944.

Goulder, R., A. S. Blanchard, P. L. Sanderson and B. Wright. A note on the recognition of pollution stress in populations of estaurine bacteria. J. Appl. Bacteriol., 46:285-289, 1979.

Graham, S., Y. Yoneda and F. E. Young. Isolation of characterization of viable deletion mutants of Bacillus subtilis Bacteriophage SP02. Gene, 7: 69-77, 1979.

Gray, B. H., C. F. Fowler, N. A. Nugent, N. Rigopoulos and R. C. Fuller. Reevaluation of Chloropseudomonas ethylica 2-K. Int. J. System. Bacteriol., 23:256-264, 1973.

Gray, T. R. G. Survival of Vegetative Microbes in Soil. In: The Survival of Vegetative Microbes, 26th Symposium of the Society for General Microbiology, T. R. G. Gray and J. R. Postgate, eds., Cambridge University Press, Cambridge, pp. 327-364, 1976.

Gray, T. R. G. and J. R. Postgate, Eds. The Survival of Vegetative Microbes. 26th Symposium of the Society for General Microbiology, Cambridge University Press, Cambridge, 1976.

Griffin, D. M. Water and Microbial Stress. In: Advances in Microbial Ecology Vol. 5, M. Alexander, ed., Plenum Publishing Corp., New York, pp. 91-136, 1981.

Grunstein, M., and D. S. Hogness. Colony hybridization: A method for the isolation of cloned DNAs that contain a specific gene. Proc. Natl. Acad. Sci. U.S., 72:3961-3965, 1975.

Gude, H. Grazing by protozoa as selection factor for activated sludge bacteria. Microbial Ecology, 5:225-237, 1979.

Gunner, H. B. and B. M. Zuckerman. Degradation of "Diazinon" by synergistic microbial action. Nature London, 217:1183-1184, 1968.

Gutenmann, W. H., J. W. Serum and D. J. Lisk. Feeding studies with VCS-438 herbicide in the dairy cow. J. Agr. Food Chem., 20:991-993, 1972.

Habte, M. and M. Alexander. Further evidence for the regulation of bacterial populations in soil by protozoa. Arch. Microbiol., 113:181-183, 1977.

Habte, M. and M. Alexander. Mechanisms of persistence of low numbers of bacteria preyed upon by protozoa. Soil Biol. Biochem., 10:1-6, 1978.

Halldal, P. Light and Microbial Activities. In: Contemporary Microbial Ecology, D. C. Ellwood, J. N. Hedger, M. J. Latham, J. M. Lynch and J. H. Slater, Eds., Academic Press, New York, pp. 1-14, 1980.

Hallewell, R. and S. Emtage. Plasmid vectors containing the tryptophan operon promoter suitable for efficient regulated expression of foreign genes. Gene, 9:27-47, 1980.

Hambrick, G. A. III, R. D. DeLaune and W. H. Patrick, Jr. Effect of estaurine sediment pH and oxidation-reduction potential on microbial hydrocarbon degradation. Appl. Environ. Microbiol., 40:365-369, 1980.

Hansen, J. B. and R. H. Olsen. Isolation of large bacterial plasmids and characterization of the P2 incompatability group plasmids pMGl and pMG5. J. Bacteriol., 135:227-238, 1978.

Harayama, S., M. Tsuda and T. Iino. High frequency mobilization of the chromosome of _Escherichia coli_ by a mutant of plasmid RP4 temperature-sensitive for maintenance. Molec. Gen. Genet., 180:47-56, 1980.

Harayama, S., M. Tsuda and T. Iino. Tnl insertion mutagenesis in _Escherichia coli_ K-12 using a temperature-sensitive mutant of plasmid RP4. Molec. Gen. Genet., 184:52-55, 1981.

Harder, W. and L. Dijkhuizen. Strategies of mixed substrate utilization in microorganisms. Phil. Trans. R. Soc. Lond. B., 297:459-480, 1982.

Harrison, D. E. F. Mixed cultures in industrial fermentation processes. Adv. Appl. Microbiol., 24:129-164, 1978.

Hartley, B. S., I. Altosaar, J. M. Dothie, and M. S. Neuberger. Experimental evolution of a Xylitol dehydrogenase. In: Structure-Function Relationships in Proteins, R. Markham and R. W. Horne, Eds., Amsterdam, North Holland, pp. 191-200, 1977.

Hartmann, J., W. Reineke and H.-J. Knackmuss. Metabolism of 3-chloro-, 4-chloro- and 3,5-dichlorobenzoate by a pseudomonad. Appl. Environ. Microbiol., 37:421-428, 1979.

Hedges, R. W. and G. A. Jacoby. Compatability and molecular properties of plasmid rms 149 in _Pseudomonas aeruginosa_ and _Escherichia coli_. Plasmid, 3:1-6, 1980.

Heffron, F., C. Rubens, and S. Falkow. Translocation of a plasmid DNA sequence which mediates ampicillin resistance: Molecular nature and specificity of insertion. Proc. Natl. Acad. Sci. U.S., 72:3623-3627, 1975.

Helling, R. B., T. Kinney and J. Adams. The maintenance of plasmid-containing organisms in populations of _Escherichia coli_. J. Gen. Microbiol., 123:129-141, 1981.

Herman, N. J. and E. Juni. Isolation and characterization of a generalized transducing bacteriophage for _Acinetobacter_. J. Virology, 13:46-52, 1974.

Hernalsteens, J.-P., F. Van Vliet, M. De Beuckeleer, A. Depicker, G. Engler, M. Lemmers, M. Holsters, M. Van Montagu and J. Schell. The Agrobacterium tumefaciens Ti plasmid as a host vector system for introducing foreign DNA in plant cells. Nature London, 287:654-656, 1980.

Heukelekian, H. and A. Heller. Relation between food concentration and surface for bacterial growth. J. Bacteriol., 40:547-558, 1940.

Hill, E. C. and A. R. Thomas. Microbiological Aspects of Supersonic Aircraft Fuel. In: Proceedings of the Third International Biodegradation Symposium, J. M. Sharpley and A. M. Kaplan, eds., Applied Science Publishers, London, pp. 157-174, 1955.

Hirsh, P., M. Bernhard, S. Cohen, J. Ensign, H. Jannasch, A. Koch, K. Marshall, A. Matin, J. Poindexter, S. Rittenberg, D. Smith and H. Veldkamp. Life Under Conditions of Low Nutrient Concentration, Group Report. In: Strategies of Microbial Life in Extreme Environments, M. Shilo, ed., Dahlem Konferenzen, Verlag Chemie, New York, pp. 357-372, 1979.

Hohn, B. and K. Murray. Packaging recombinant DNA molecules into bacteriophage particles in vitro. Proc. Nat'l. Acad. Sci. U.S., 74:3259-3263, 1977.

Holladay, D. W., C. W. Hancher, C. D. Scott and D. D. Chilcote. Biodegradation of phenolic waste liquors in stirred-tank, packed-bed and fluidized-bed bioreactors. J. Water Poll. Contr. Fed., 50:2573-2589, 1978.

Holloway, B. W., M. H. Richmond. R.-factors used for genetic studies in strains of Pseudomonas aeruginosa and their origin. Genet. Res. Camb., 21:103-105, 1973.

Holloway, B. W., V. Krishnapillai and V. Stanisish. Pseudomonas genetics. Ann. Rev. Genet., 5:425-446, 1971.

Hooykaas, P. J. J., H. den Dulk-Ras, R. A. Schilperoort. Molecular mechanism of Ti plasmid mobilization by R plasmids: Isolation of Ti plasmids with transposon-insertions in Agrobacterium tumefaciens. Plasmid, 4:64-75, 1980.

Hopper, D. J. and P. D. Kemp. Regulation of enzymes of 3,5-xylenol-degradative pathway in Pseudomonas putida: Evidence for a plasmid. J. Bacteriol., 142:21-26, 1980.

Hopwood, D. A. Genetic studies with bacterial protoplasts. Ann. Rev. Microbiol., 35:237-272, 1981.

Hopwood, D. A. and H. M. Wright. Bacterial protoplast fusion: Recombination in fused protoplasts of Streptomyces coelicolor. Molec. Gen. Genet., 162: 307-317, 1978.

Hopwood, D. A. and H. M. Wright. Factors affecting recombinant frequency in protoplast fusions of Streptomyces coelicolor. J. Gen. Microbiol., 111:137-143, 1979.

Horikoshi, K. and T. Akiba. Alkalophilic Microorganisms: a New Microbial World, Springer Verlag, New York, 1982.

Horinouchi, S., T. Uozumi and T. Beppu. Cloning of Streptomyces DNA into Escherichia coli: Absence of heterospecific gene expression of Streptomyces genes in E. coli. Agr. Biol. Chem., 44:367-381, 1980.

Horvath, R. S. Cometabolism of the herbicide 2,3,6-trichlorobenzoate. J. Agr. Food Chem., 19:291-293, 1971.

Horvath, R. S. Microbial co-metabolism and the degradation of organic compounds in nature. Bacteriol. Rev., 36:146-155, 1972.

Horvath, R. S. and M. Alexander. Cometabolism: A technique for the accumulation of biochemical products. Can. J. Microbiol., 16:1131-1132, 1970a.

Horvath, R. S. and M. Alexander. Co-metabolism of M-chlorobenzoate by an Arthrobacter. Appl. Microbiol., 20:254-258, 1970b.

Hsia, M. T. S., F. V. Z. Bairstow, L. C. T. Shih, J. G. Pounds and J. R. Allen. 3,4,3',4'-Tetrachloroazobenzene: A potential environmental toxicant. Res. Commun. Chem. Pathol. Pharmacol., 17:225-236, 1977.

Hsu, T.-S. and R. Bartha. Accelerated mineralization of two organophosphate insecticides in the rhizosphere. Appl. Environ. Microbiol., 37:36-41, 1979.

Hulbert, M. H. and S. Krawiec. Cometabolism: a critique. J. Theor. Biol., 69:287-291, 1977.

Hutchison, K. W., and H. O. Halvorson. Cloning of randomly sheared DNA fragments from a phi-105 lysogen of Bacillus subtilis identification of prophage-containing clones. Gene, 8:267-278, 1980.

Jackson, D. A., R. H. Symons and P. Berg. Biochemical method for inserting new genetic information into DNA of simian virus 40: circular SV40 DNA molecules containing lambda phage genes and the galactose operon of Escherichia coli. Proc. Nat'l Acad. Sci. U.S., 69:2904-2909, 1972.

Jacoby, G. A. and J. A. Shapiro. Plasmids Studied in Pseudomonas aeruginosa and Other Pseudomonads. In: DNA Insertion Elements, Plasmids and Episomes, A. I. Bukhari, J. A. Shapiro and S. L. Adhya, eds., Cold Harbor Laboratory, New York, pp. 639-656, 1977.

Jamison, V. W., R. L. Raymond and J. O. Hudson. Microbial hydrocarbon co-oxidation. III. isolation and characterization of an α,-α'-dimethyl-cis, cis-muconic acid - producing strain of Nocardia corallina Appl. Microbiol., 17:853-856, 1969.

Jamison, V. W., R. L. Raymond and J. O. Hudson. Hydrocarbon co-oxidation by Nocardia corallina V-49. Developments in Industrial Microbiol. 12:99-105, 1971.

Jannasch, H. W. Microbial Ecology of Aquatic Low Nutrient Habitats. In: Strategies of Microbial Life in Extreme Environments, M. Shilo, ed., Dahlem Konferenzen, Verlag Chemie, New York, pp. 245-260, 1979.

Jannasch, H. W. and P. H. Pritchard. The role of inert particulate matter in the activity of aquatic microorganisms. Mem. Inst. Ital. Idrobiol., 29(Suppl.):289-308, 1972.

Jannasch, H. W., K. Eimhjellen, C. O. Wirsen and A. Farmanfarmaian. Microbial degradation of organic matter in the deep sea. Science, 171:672-675, 1971.

Jensen, H. L. Decomposition of chloro-substituted aliphatic acids by soil bacteria. Can. J. Microbiol., 3:151-164, 1957.

Jensen, H. L. Decomposition of chlorine-substituted organic acids by fungi. Acta Agr. Scand., 9:421-434, 1959.

Jensen, H. L. Decomposition of chloroacetates and chloropropionates by bacteria. Acta Agr. Scand., 10:83-103, 1960.

Jobson, A., F. D. Cook, and W. S. Westlake. Microbial utilization of crude oil. Appl. Microbiol., 23:1082-1089, 1972.

Johanides, V. and D. Hrsak. Changes in mixed bacterial culture during linear alkylbenzenesulfonate biodegradation. Proc. 5th Internat. Fermentation Symp., Berlin, Abst. 23.08, 1976.

Johnston, A. W.B., J. L. Beynon, A. V. Buchanan-Wollaston, S. M. Setchell, P. R. Hirsch, and J. E. Beringer. High frequency transfer of nodulating ability between strains and species of *Rhizobium*. Nature, 276:634-636, 1978.

Johnston, J. B. and I. C. Gunsalus. Isolation of Metabolic plasmid DNA from *Pseudomonas putida*. Biochem. Biophys. Res. Commun., 75:13-19, 1977.

Johnston, J. B. and S. G. Robinson. The Development of New Pollution Control Technologies Using Genetic Engineering Methods -- An Assessment of Problems and Opportunities. In: Proceedings of the Battelle International Conference on Genetic Engineering, Vol. IV, Battelle, Seattle, pp. 184-203, 1981.

Johnston, J. B. and S. G. Robinson. Opportunities for the Development of New Detoxication Processes Through Genetic Engineering. In: Proceedings of the ACS Symposium on Detoxification of Hazardous Chemicals, J. Exner, Ed., Ann Arbor Science Press, Ann Arbor, pp. 301-314, 1982a.

Johnston, J. B. and S. G. Robinson. Development of New Pollution Control Technologies: Opportunities and Problems. In: Proceedings of the Notre Dame/Hooker Symposium on Genetic Engineering Aspects of Waste Treatment. In the press, 14 pp., 1982b.

Jones, R. D. and M. A. Hood. Interaction between an ammonium-oxidizer, Nitrosomonas sp. and two heterotrophic bacteria, Nocardia atlantica and Pseudomonas sp., a note. Microbial Ecology, 6:271-275, 1980.

Juengst, F. W. and M. Alexander. Conversion of 1,1,1-trichloro-2,2-bis[p-chlorophenyl)ethane (DDT) to water soluble products by microorganisms. J. Agr. Food Chem., 24:111-115, 1976.

Juni, E. Interspecies transformation of Acinetobacter: Genetic evidence for a ubiquitous genus. J. Bacteriol., 112:917-931, 1972.

Juni, E. Simple genetic transformation assay for rapid diagnosis of Moraxella osloensis. Appl. Microbiol., 27:16-24, 1974.

Juni, E. Genetics transformation assays for identification of strains of Moraxella urethralis. J. Clin. Microbiol., 5:227-235, 1977.

Juni, E. Genetics and physiology of Acinetobacter. Ann Rev. Microbiol., 32:349-371, 1978.

Juni, E. and G. A. Heym. Transformation assay for identification of psychrotophic Achromobacters. Appl. Environ. Microbiol., 40:1106-1114, 1980.

Juni, E. and A. Janik. Transformation of Acinetobacter calco-aceticus (Bacterium anitratum). J. Bacteriol., 98:281-288, 1969.

Kachholz, T. and H. J. Rehm. Degradation of long chain alkanes by bacilli II metabolic pathways. Eur. J. Appl. Microbiol. Biotechnol., 6: 39-54, 1978.

Kachholz, T. and H. J. Rehm. Degradation of long chain alkanes by bacilli Eur. J. Appl. Microbiol. Biotechnol., 10:95-97, 1980.

Kallman, B. J. and A. K. Andrews. Reductive dechlorination of DDT to DDD by yeast. Science, 141:1050-1051, 1963.

Kamp, P. F. and A. M. Chakrabarty. Plasmids Specifying p-chlorobiphenyl Degradation in Enteric Bacteria. In: Plasmids of Medical, Environmental and Commercial Importance, K. Timmis and A. Puhler, eds., Elsevier, Amsterdam, pp. 275-285, 1979.

Kargi, F. and J. M. Robinson. Microbial desulfurization of coal by thermophilic microorganisms Sulfolobus acidocaldarius. Biotech. Bioeng., 24:2115-2121, 1982a.

Kargi, F. and J. M. Robinson. Removal of sulfur compounds from coal by the thermophilic organism Sulfolobus acidocaldarius. Appl. Environ. Microbiol., 44:878-883, 1982b.

Karl, D. M., C. O. Wirsen and H. W. Jannasch. Deep-sea primary production at the Galapogos hydrothermal vents. Science, 207:1345-1347, 1980.

Kawasaki, H., N. Tone and K. Tonomura. Plasmid-determined dehalogenation of haloacetates in Moraxella species. Agr. Biol. Chem., 45:29-34, 1981a.

Kawasaki, H., H. Yahara and K. Tonomura. Isolation and characterization of plasmid pUO1 mediating dehalogenation of haloacetate and mercury resistance in Moraxella sp. B. Agr. Biol. Chem., 45:1477-1481, 1981b.

Kawasaki, H., N. Tone and K. Tonomura. Purification and properties of haloacetate halidohydrolase specified by plasmid from Moraxella sp. strain B. Agr. Biol. Chem., 45:35-42, 1981c.

Kellogg, S. T., D. K. Chatterjee and A. M. Chakrabarty. Plasmid assisted molecular breeding: New technique for enhanced biodegradation of persistent toxic chemicals. Science, 214:1133-1135, 1981.

Kemp, D. J. and A. F. Cowman. Direct immunoassay for detecting E. coli colonies that contain polypeptides encloded by cloned DNA segments. Proc. Nat'l. Acad. Sci. U.S., 78:4520-4524, 1981.

Kilbane, J. J., D. K. Chatterjee, J. S. Karns, S. T. Kellogg and A. M. Chakrabarty. Biodegradation of 2,4,5-trichlorophenoxyacetic acid by a pure culture of Pseudomonas cepacia. Appl. Environ. Microbiol., 44:72-78, 1982.

Kimbrough, R. D., Ed. Halogenated Biphenyls, Terphenyls, Naphthalenes, Dibenzodioxins and Related Products, Elsevier, New York, 1980.

Kiyohara, H., K. Nagao, K. Kouno and K. Yano. Phenanthrene-degrading phenotype of Alcaligenes faecalis AFK2. Appl. Environ. Microbiol., 43: 458-461, 1982.

Kleckner, N., J. Roth and D. Botstein. Genetic engineering in vivo using translocatable drug-resistant elements: New methods in bacterial genetics. J. Mol. Biol., 116:125-159, 1977.

Knackmuss, H.-J. Degradation of Halogenated and Sulfonated Hydrocarbons. In: Microbial Degradation of Xenobiotics and Recalcitrant Compounds, T. Leisinger, A. M. Cook, R. Hutter and J. Nuesch, eds., Academic Press, New York, pp. 189-212, 1981.

Knackmuss, H.-J. and M. Hellwig. Utilization and cooxidation of chlorinated phenols by Pseudomonas sp. B13. Arch. Microbiol., 117:1-7, 1978.

Kordyum, V. A., N. A. Kozyrovskaya, R. I. Gvozdyak and V. A. Muras. Transfer of plasmid RP41 markers into Xanthomonas beticola. Akad. Nauk Ukr. S.S.R., Kiev. Dopovidi Ser. B: Geol. Khim. Biol. Nauki 0(5):78-80; english translation pp. 82-85, 1980.

Konicek, J. and M. Konickova-Radochova. Possibilities of the conjugation process in Mycobacteria. Folia Microbiol., 20:382-388, 1975.

Kostriken, R., C. Morita and F. Heffron. Transposon Tn 3 encodes a site-specific recombination system: Identification of essential sequences, genes, and actual site of recombination. Proc. Nat'l. Acad. Sci. U.S., 78:4041-4045, 1981.

Kreis, M., J. Eberspacher and F. Lingens. Detection and characterization of plasmids in Chloridazon and antipyrin degrading bacteria. Zbl. Bakt. Hyg. I. Abt. Orig., C2:45-60, 1981.

Krugel, H., G. Fiedler and D. Noack. Transfection of protoplasts from Streptomyces lividans 66 with Actinophage SH10 DNA. Mol. Gen. Genet., 177:297-300, 1980.

Kuenen, J. G., J. Boonstra, H. Schroder and H. Veldkamp. Competition for inorganic substrates among chemoorganotrophic and chemolithotrophic bacteria. Microbial Ecology, 3:119-130, 1977.

LaLiberte, P. and D. J. Grimes. Survival of Escherichia coli in lake bottom sediments. Appl. Environ. Microbiol., 43:623-628, 1982.

Larson, R. J. and J. L. Pate. Glucose transport in isolated prosthecae of Asticcacaulis biprosthecum. J. Bacteriol., 126:282-293, 1976.

Lawton, W. D., B. C. Morris and T. W. Burrows. Gene transfer in strains of Pasteurella pseudotuberculosis. J. Gen. Microbiol., 52:25-34, 1968.

Leadbetter, E. R. and J. W. Foster. Oxidation products formed from gaseous alkanes by the bacterium Pseudomonas methanica. Arch. Biochem. Biophys., 82:491-492, 1959.

LeBlanc, D. J. and R. P. Mortlock. Metabolism of D-arabinose: Origin of a D-ribulokinase activity in Escherichia coli. J. Bacteriol., 106:82-89, 1971a.

LeBlanc, D. J. and R. P. Mortlock. Metabolism of D-arabinose: A new pathway in Escherichia coli. J. Bacteriol., 106:90-96, 1971b.

Levin, B. R. and V. A. Rice. The kinetics of transfer of nonconjugative plasmids by mobilizing conjugative factors. Genet. Res., Camb., 35:241-259, 1980.

Levin, B. R. and F. M. Stewart. The population biology of bacterial plasmids: A Priori conditions for the existence of mobilizable nonconjugative factors. Genetics, 94:425-443, 1980.

Levin, B. R., F. M. Stewart and V. A. Rice. The kinetics of conjugative plasmid transmission: Fit of a simple mass action model. Plasmid, 2:247-260, 1979.

Levy, S. B. and B. Marshall. Survival of E. coli host-vector systems in the human intestinal tract. Recombinant DNA technical bulletin, National Institute of Health, Washington, DC, 2(2):77-80, 1979.

Levy, S. B., B. Marshall, D. Rowse-Eagle and A. Onderdonk. Survivial of Escherichia coli host-vector systems in the mammalian intestine. Science, 209:391-394, 1980.

Li, A. W., P. J. Krell, and D. E. Mahony. Plasmid detection in a bacteriocinogenic strain of Clostridium perfringens. Can. J. Microbiol., 26:1018-1022, 1980.

Little, P., P. Curtis, Ch. Coutelle, J. Van Den Berg, R. Dalgleish, S. Malcolm, M. Courtney, P. Westaway, and R. Williamson. Isolation and partial sequence of recombinant plasmids containing human α-, β- and γ-globin cDNA fragments. Nature London, 273:640-643, 1978.

Liu, D. Enhancement of PCB s biodegradation by sodium ligninsulfonate. Water Res., 14:1467-1475, 1980.

Lobban, P. E. and A. D. Kaiser. Enzymatic end-to-end joining of DNA molecules. J. Mol. Biol., 78:453-471, 1973.

Loenen, W. A. M. and W. J. Brammar. A bacteriophage lambda vector for cloning large DNA fragments made with several restriction enzymes. Gene, 10:249- 259 1980.

Low, K. B. and D. D. Porter. Modes of gene transfer and recombination in bacteria. Ann. Rev. Genet., 12:249-287, 1978.

Lowendorf, H. S., A. M. Baya and M. Alexander. Survival of Rhizobium in acid soils. Appl. Environ. Microbiol., 42:951-957, 1981.

Lukins, H. B. and J. W. Foster. Utilization of hydrocarbons and hydrogen by mycobacteria. Z. Allg. Mikrobiol., 3:251-264, 1963.

Lynch, J. M. Interactions Between Bacteria and Plants in the Root Environment. In: Bacteria and Plants, M. E. Rhodes-Roberts and F. A. Skinner, Eds., Academic Press, New York, pp. 1-23, 1982.

Lyons, C. D. and J. Savage. Cometabolism of lindane by bacterial isolates. Abstr. Ann. Meeting Am. Soc. Microbiol., pp. 232, 1979.

Mach, P. A. and D. J. Grimes. R-plasmid transfer in a wastewater treatment plant. Appl. Environ. Microbiol., 44:1395-1403, 1982.

MacRae, I. C., K. Raghu and E. M. Bautista. Anaerobic degradation of the insecticide lindane by Clostridium sp. Nature London, 221:859-860, 1969.

Madsen, E. L. and M. Alexander. Transport of Rhizobium and Pseudomonas through soil. Soil Sci. Soc. Am. J., 46:557-560, 1982.

Magrum, L. J., K. W. Luehrsen and C. R. Woese. Are extreme halophiles actually "bacteria." J. Mol. Evol., 11:1-8, 1978.

Marshall, B., S. Schluederberg, C. Tachibana and S. B. Levy. Survival and transfer in the human gut of poorly mobilizable (pBR322) and of transferable plasmids from the same carrier E. coli. Gene, 14:145-154, 1981.

Matney, T. S. and N. E. Achenback. A comment on the fertility of F_2 donor types of Escherichia coli K12. Biochem. Biophys. Res. Commun., 9:285-287, 1962.

Matsumura, F. and H. J. Benezet. Studies on the bioaccumulation and microbial degradation of 2,3,7,8-tetrachlorodibenzo-p-dioxin. Environ. Health Perspect., 5:253-258, 1973.

Matsumura, F. and G. M. Boush. Degradation of insecticides by a soil fungus, Trichoderma viride. J. Econ. Entomol., 61:610-612, 1968.

McConnell, E. E. and J. A. Moore. The Toxicopathology of TCDD. In: Dioxin: Toxicological and Chemical Aspects, F. Cattabeni, A. Cavallaro and G. Galli, eds., John Wiley and Sons, New York, pp. 137-142, 1978.

McElroy, M. B., J. W. Elkins, S. C. Wofsy and Y. L. Yung. Sources and sinks for atmospheric N_2O. Reviews of Geophysics and Space Physics, 14:143-150, 1976.

McKinney, J. and E. McConnell. Structural Specificity and the Dioxin Receptor. In: Chlorinated Dioxins and Related Compounds, Impact on the Environment, O. Hutzinger, R. W. Frei, E. Merian and F. Pocchiari, eds., Pergamon Press, New York, pp. 367-381, 1982.

McLaren, A. D. and J. Skujins. The Physical Environment of Microorganisms in Soil. In: The Ecology of Soil Bacteria, T. R. G. Gray and D. Parkinson, eds., Liverpool University Press, Liverpool, pp. 3-24, 1968.

Megee, R. D., J. F. Drake, A. G. Fredrickson and H. M. Tsuchiya. Studies in intermicrobial symbiosis: S. cerevisiae and L. casei. Can. J. Microbiol., 18:1733-1742, 1972.

Mendez-Castro, F. A. and M. Alexander. Acclimation of Rhizobium to salts, increasing temperature and activity. Rev. Latinoam. Microbiol., 18:155-158, 1976.

Merkel, G. J. and J. J. Perry. Increased cooxidative biodegradation of Malathion in soil via cosubstrate enrichment, Agr. Food Chem., 25:1011-1012, 1977.

Miles, J. R. W., C. M. Tu and C. R. Harris. Degradation of Heptachlor epoxide and Heptachlor by a mixed culture of soil microorganisms. J. Econ. Entomol., 64:839-841, 1971.

Miller, R. V., J. M. Pemberton and K. E. Richards. F116, D3 and G101: Temperate bacteriophages of Pseudomonas aeruginosa. Virology, 59: 566-569 1974.

Monod, J. Recherches sur la Croissance des Cultures Bactériennes, Hermann, Paris, 1942.

Morris, C. M. and E. A. Barnsley. The cometabolism of 1- and 2-chloronaphthalene by pseudomonads. Can. J. Microbiol., 28:73-79, 1982.

Mortlock, R. P. Metabolic acquisitions through laboratory selection. Ann Rev. Microbiol., 36:259-284, 1982.

Motosugi, K., N. Esaki and K. Soda. Purification and properties of 2-halo acid dehalogenase from Pseudomonas putida. Agr. Biol. Chem., 46:837-838, 1982.

Mulkins-Phillips, G. J. and J. E. Stewart. Distribution of hydrocarbon utilizing bacteria in Northwestern Atlantic waters and coastal sediments. Can. J. Microbiol., 20:955-962, 1974.

Murray, K. Genetic Engineering: Possibilities and prospects for its applications in industrial microbiology. Phil. Trans. R. Soc. London B., 290:369-386, 1980.

Nakano, E. and T. Masuda. Construction of "sleeper" cloning vehicles for enzyme overproduction. Agr. Biol. Chem., 46:313-315, 1982.

Nathans, D. and H. O. Smith. Restriction endonucleases in the analysis and restructuring of DNA molecules. Ann Rev. Biochem., 44:273-293, 1975.

National Research Council. Causes and Effects of Stratospheric Ozone Reduction: An Update, National Academy Press, Washington, DC, pp. 187-192, 1982.

Novakova, J. Effect of increasing concentrations of clays on the decomposition of glucose. I. effect of bentonite. Zentralb. Bakt. Parasit. Infekt. Hyg. II Abt. II, 127:359-366, 1972a.

Novakova, J. Effect of increasing concentrations of clay on the decomposition of glucose. II. effect of kaolinite. Zentralb. Bakt. Parasit. Infekt. Hyg. II Abt. II, 127:367-372 1972b.

Okanishi, M. and Y. Okami. Isolation and characteristics of an actinophage active on Streptomyces kanamyceticus. J. Gen. Appl. Microbiol., 12:207-217, 1966.

Okanishi, M., R. Utahara and Y. Okami. Infection of the protoplasts of Streptomyces kanamyceticus with deoxyribonucleic acid preparation from actinophage PK-66. J. Bacteriol., 92:1850-1852, 1966.

Okanishi, M., K. Hamana and H. Umezawa. Factors affecting infection of protoplasts with deoxyribonucleic acid of actinophage PK-66. J. Virol. 2:686-691, 1968.

Oliver, E. J. and R. P. Mortlock. Growth of Aerobacter aerogenes on D-arabinose: origin of the enzyme activities. J. Bacteriol., 108:287-292, 1971a.

Oliver, E. J. and R. P. Mortlock. Metabolism of D-arabinose by Aerobacter aerogenes: Purification of the isomerase. J. Bacteriol., 108:293-299, 1971b.

Omori, T. and M. Alexander. Bacterial dehalogenation of halogenated alkanes and fatty acids. Appl. Environ. Microbiol., 35:867-871, 1978.

Orndorff, S. A. and R. R. Colwell. Microbial transformation of kepone. Appl. Environ. Microbiol., 39:398-406, 1980.

Osa-Afiana, L. O. and M. Alexander. Effect of moisture on the survival of Rhizobium in soil. Soil Sci. Soc. Am. J., 43:925-930, 1979.

Ou, L. T. and H. C. Sikka. Extensive degradation of Silvex by synergistic action of aquatic microorganisms. J. Agr. Food Chem., 25:1336-1339, 1977.

Palchaudhuri, S. Molecular characterization of hydrocarbon degradative plasmids in Pseudomonas putida. Biochem. Biophys. Res. Commun., 77:518-525, 1977.

Palm, J. C., D. Jenkins and D. S. Parker. Relationship between organic loading, dissolved oxygen concentration and sludge settleabilitv in the completely-mixed activated sludge process. J. Water Poll. Control Fed., 52:2484-2506, 1980.

Paris, D. F., D. L. Lewis and J. T. Barnett. Bioconcentration of toxaphene by microorganisms. Bull. Environ. Contam. Toxicol., 17:564-571, 1977.

Pearce, B. A. and M. T. Heydeman. Metabolism of di(ethylene glycol) (2-(2'-Hydroxyethoxy)ethanol) and other short poly(ethylene glycol)s by Gram negative bacteria. J. Gen. Microbiol., 118:21-27, 1980.

Pemberton, J. M. and P. R. Fisher. 2,4-D plasmids and persistence. Nature London, 268:732-733, 1977.

Pereira, M. R. and M. A. Benjaminson. Broadcast of microbial aerosols by stacks of sewage treatment plants and effects of ozonation on bacteria in the gaseous effluent. Public Health Reports, 90:208-212, 1975.

Perry, J. J. Microbial cooxidations involving hydrocarbons. Microbiol. Rev., 43:59-72, 1979.

Pfaender, F. K. and M. Alexander. Extensive microbial degradation of DDT in vitro and DDT metabolism by natural communities. J. Agr. Food Chem., 20:842-846, 1972.

Philippi, M., V. Krasnobajew, J. Zeyer and R. Hutter. Fate of TCDD in Microbial Cultures and in Soil Under Laboratory Conditions. In: Microbial Degradation of Xenobiotics and Recalcitrant Compounds. FEMS Symp. no. 12, T. Leisinger, R. Hutter, A. M. Cook and J. Nuesch, eds., Academic Press, New York, pp. 221-236, 1981.

Pirt, S. J. Principles of Microbe and Cell Cultivation, John Wiley and Sons, New York, pp. 63-75, 1975.

Plimmer, J. R., P. C. Kearney and D. W. von Endt. Mechanism of conversion of DDT to DDD by _Aerobacter aerogenes_. J. Agr. Food Chem., 16:594-597, 1968.

Poindexter, J. S. The caulobacters: ubiquitous unusual bacteria. Microbiol. Rev., 45:123-179, 1981a.

Poindexter, J. S. Oligotrophy, Fast and Famine Existence. In: Advances in Microbial Ecology, vol. 5, M. Alexander, ed., Plenum Press, New York, pp. 63-89, 1981b.

Poland, A., W. F. Greenlee and A. S. Kende. Studies on the Mechanism of Action of the Chlorinated Dibenzo-p-dioxins and Related Compounds. Health Effects of Halogenated Aromatic Hydrocarbons, W. J. Nicholson and J. A. Moore, eds., Annals of the N.Y. Acad. Sci. vol. 320, New York Academy of Science, New York, pp. 214-230, 1979.

Ramakrishnan, T. and M. S. Shaila. Interfamilial transfer of amber suppressor gene for the isolation of amber mutants of mycobacteriophage I3. Arch. Microbiol., 120:301-302, 1979.

Ramirez, C. and M. Alexander. Evidence suggesting protozoan predation on _Rhizobium_ associated with germinating seeds and in the rhizosphere of beans (_Phaseolus vulgaris_ L.). Appl. Environ. Microbiol., 40:492-499, 1980.

Ratledge, C. and F. G. Winder. Biosynthesis and utilization of aromatic compounds by _Mycobacterium smegmatics_ with particular reference to the origin of salicylic acid. Biochem. J., 101:274-283, 1966.

Raymond, R. L., V. W. Jamison and J. O. Hudson. Microbial hydrocarbon co-oxidation. I. oxidation of mono- and dicyclic hydrocarbons by soil isolates of the genus _Norcardia_. Appl. Microbiol., 15:857-865, 1967.

Reich, M. and R. Bartha. Degradation and mineralization of a polybutene film-mulch by the synergistic action of sunlight and soil microbes. Soil Sci., 124:177-180, 1977.

Reid, D. S. Water Activity as the Criterion of Water Availability. In: Contemporary Microbial Ecology, D. C. Ellwood, J. N. Hedger, M. J. Latham, J. M. Lynch and J. H. Slater, eds., Academic Press, New York, pp. 15-28, 1980.

Reineke, W. and H.-J. Knackmuss. Chemical structure and biodegradability of halogenated aromatic compounds. Substituent effects on 1,2-dioxygenation of benzoic acid. Biochim. Biophys. Acta, 542:412-423, 1978a.

Reineke, W. and H.-J. Knackmuss. Chemical structure and biodegradability of halogenated aromatic compounds. Substituent effect on dehydrogenation of 3,5-cyclohexadiene-1,2-diol-1-carboxylic acid. Biochim. Biophys. Acta, 542:424-429, 1978b.

Reineke, W. and H.-J. Knackmuss. Construction of haloaromatics utilizing bacteria. Nature London 277:385-386, 1979.

Reineke, W., S. W. Wessels, M. A. Rubio, J. Latorre, U. Schwien, E. Schmidt, M. Schlomann and H.-J. Knackmuss. Degradation of Monochlorinated aromatics following transfer of genes encoding chlorocatechol catabolism. FEMS Microbiol. Lett., 14:291-294, 1982.

Reiner, J. M. Behavior of Enzyme Systems, Burgess, Minneapolis, p. 2, 1959.

Reutergardh, L. Chlorinated Hydrocarbons in Estuaries. In: Chemistry and Biogeochemistry of Estuaries, F. Olausson and I. Cato., eds., John Wiley & Sons, New York, pp. 349-365.

Rheinwald, J. C., A. M. Chakrabarty and I. C. Gunsalus. A transmissible plasmid controlling camphor oxidation in _Pseudomonas putida_. Proc. Nat'l Acad. Sci. U.S., 70:885-889, 1973.

Riding, J. T., W. R. Elliott and J. H. Sherrard. Activated sludge phosphorus removal mechanisms. J. Wat. Poll. Control Fed., 51:1040-1053, 1979.

Roberts, T. M., R. Kacich and M. Ptashne. A general method for maximizing the expression of a cloned gene. Proc. Nat'l. Acad. Sci. U.S., 76; 760-764, 1979.

Roberts, T. M., S. L. Swanberg, A. Poteete, G. Riedel and K. Backman. plasmid cloning vehicle allowing a positive selection for inserted fragments. Gene, 12:123-127, 1980.

Robinson, M. K., P. M. Bennett and M. H. Richmond. Inhibition of TnA translocation by TnA. J. Bacteriol., 129:407-414, 1977.

Robinson, M. K., P. M. Bennett, S. Falkow, and H. M. Dodd. Isolation of a temperature-sensitive derivative of RP1. Plasmid, 3:343-347, 1980.

Rodicio, M. R. and K. F. Chater. Small DNA-Free liposomes stimulate transfection of _Streptomyces_ protoplasts. J. Bacteriol., 151:1078-1085, 1982.

Rogoff, M. R. Oxidation of aromatic compounds by bacteria. Adv. Appl. Microbiol., 3:193-221, 1961.

Sagai, H., S. Iyobe and S. Mitsuhashi. Inhibition and facilitation of transfer among Pseudomonas aeruginosa R plasmids. J. Bacteriol., 131: 765-769, 1977.

Sagik, B. P. and C. A. Sorber. The survival of host-vector systems in domestic sewage treatment plants. Recombinant DNA technical bulletin, National Institute of Health, Washington, DC, 2(2):55-61, 1979.

Sagoo, G. S. and R. B. Cain. Factors affecting the transfer of the catabolic plasmid spedifying the utilization of alkylbenzene sulphonates between species of Pseudomonas. Abst. Soc. Gen. Microbiol. Quarterly, 6:17, 1979.

Sakazawa, C., M. Shimao, Y. Taniguchi, and N. Kato. Symbiotic utilization of polyvinyl alcohol by mixed cultures. Appl. Environ. Microbiol., 41:261-267, 1981.

Salkinoja-Salonen, M. S. and V. Sundman. Regulation and Genetics of the Biodegradation of Lignin Derivatives in Pulp Mill Effluents. In: Lignin Biodegradation: Microbiology, Chemistry and Potential Applications, T. K. Kirk, ed., CRC Press, Cleveland, OH, pp. 179-198, 1979.

Salkinoja-Salonen, M. S., E. Vaisanen and A. Paterson. Involvement of Plasmids in the Bacterial Degradation of Lignin-derived Compounds. In: Plasmids of Medical, Environmental and Commercial Importance, K. N. Timmis, and A. Puhler, eds., Elsevier, Amsterdam, pp. 301-314, 1979.

Sanderson, K. E., H. Ross, L. Ziegler and P. H. Makela. F^+, Hfr and F' strains of Salmonella typhimurium and Salmonella abony. Bacteriol. Rev., 36:608-637, 1972.

Sawula, R. V. and I. P. Crawford. Mapping of the tryptophan genes of Acinetobacter calcoaceticus by transformation. J. Bacteriol., 112: 797-805, 1972.

Sayler, G. S., M. P. Shiaris, W. Beck and S. Held. Effects of polychlorinated biphenyls and environmental biotransformation products on aquatic nitrification. Appl. Environ. Microbiol., 43:949-952, 1982.

Schaeffer, P., B. Cami and R. D. Hotchkiss. Fusion of Bacterial Protoplasts. Proc. Nat'l. Acad. Sci. U.S., 73:2151-2155, 1976.

Scheller, R. H., R. E. Dickerson, H. W. Boyer, A. D. Riggs and K. Itakura. Chemical synthesis of restriction enzymes recognition sites useful for cloning. Science, 196:177-180, 1977.

Schroth, M. N. and J. G. Hancock. Disease-suppressive soil and root-colonizing bacteria. Science, 216:1376-1381, 1982.

Schuphan, I. and K. Ballschmiter. Metabolism of polychlorinated norbornenes by Clostridium butyricum. Nature London, 237:100-101, 1972.

Schupp, T., R. Hutter and D. A. Hopwood. Genetic recomination in Norcardia mediterranei. J. Bacteriol., 121:128-136, 1975.

Senior, E., A. T. Bull and J. H. Slater. Enzyme evolution in a microbial community growing on the herbicide Dalapon. Nature London, 263: 476-479, 1976.

Sermonti, G., L. Lanfaloni and M. R. Micheli. A jumping gene in Streptomyces coelicolor A3(2). Mol. Gen. Genet., 177:453-458, 1980.

Sevastopoulos, G. C., C. T. Wehr and D. A. Glaser. Large scale automated isolation of Escherichia coli mutants with temperature-sensitive DNA replication. Proc. Nat'l. Acad. Sci. U.S., 74:3485-3489, 1977.

Sezgin, M. The role of filamentous microorganisms in activated sludge settling. Prog. Water. Tech., 12:97-107, 1980.

Sgaramella, V. Enzymatic oligomerization of bacteriphage P22 DNA and of linear simian virus 40 DNA. Proc. Nat'l. Acad. Sci. U.S., 69: 3389-3393, 1972.

Shapiro, J. A., S. L. Adhya and A. I. Bukhari. Introduction: New Pathways in the Evolution of Chromosome Structure. In: DNA Insertion Elements, Plasmids and Episomes, A. I. Bukhari, J. A. Shapiro and S. L. Adhya, eds., Cold Spring Harbor Laboratory, New York, pp. 3-12, 1977.

Sharpe, A. N., D. S. Clark and A. Balows. Mechanizing Microbiology, Chas. C. Thomas, Springfield, IL, 1978.

Shivvers, D. W. and T. D. Brock. Oxidation of elemental sulfur by Sulfolobus acidocaldarius. J. Bacteriol., 114:706-710, 1973.

Shoda, M., T. Ohsumi and S. Udaka. A highly phosphate accumlating bacterium: A candidate for phosphate removal from wastewater. Abstr. of Ann. Meeting of Am. Soc. Microbiol., p. 237, 1979.

Shortle, D., D. DiMaio and D. Nathans. Directed Mutagenesis. Ann. Rev. Genet., 15:265-294, 1981.

Skalka, A. and L. Shapiro. In situ immunoassays for gene translation products in phage plaques and bacterial colonies. Gene, 1:65-79, 1976.

Skryabin, G. K., V. V. Kochetkov, A. A. Eremin, A. N. Perebityuk, I. I. Starovoitov and A. M. Boronin. New naphthalene-biodegrading plasmid pBS4. Dokl. Akad. Nauk. SSSR, 250:212-215, 1980.

Slater, J. H. and A. T. Bull. Interactions Between Microbial Populations. In Companion to Microbiology, A. T. Bull and P. M. Meadow, Eds., Longman, New York, pp 181-206, 1978.

Slater, J. H. and A. T. Bull. Environmental microbiology: Biodegradation. Phil. Trans. R. Soc. Lond. B., 297:575-597, 1982.

Slater, J. H., D. Lovatt, A. J. Weightman, E. Senior and A. T. Bull. The growth of Pseudomonas putida on chlorinated aliphatic acids and its dehalogenase activity. J. Gen. Microbiol., 114:125-136, 1979.

Smith, M. and S. Gillam. In Vitro Construction of Specific Mutants. In: Proceedings of the ICN-UCLA Symposium on Developmental Biology Using Purified Genes, Vol. XXIII, pp. 671-682, 1981a.

Smith, M. and S. Gillam. Constructed Mutants Using Synthetic Oligodeoxyribonucleotides as Site-specific Mutagens. In: Genetic Engineering, Vol. 3, J. K. Setlow and A. Hollaender, eds., Plenum, New York, pp. 1-32, 1981b.

Stanisich, V. A. and M. H. Richmond. Gene Transfer in the Genus Pseudomonas. In: Genetics and Biochemistry of Pseudomonas, P. H. Clarke, ed., John Wiley and Sons, New York, pp. 163-190, 1975.

Stanlake, G. J. and R. K. Finn. Isolation and characterization of a pentachlorophenol-degrading bacterium. Appl. Environ. Microbiol, 44:1421-1427, 1982.

Steen, H. B. and E. Boye. Escherichia coli growth studies by dual-parameter flow cytophotometry. J. Bacteriol., 145:1091-1094, 1980.

Stevenson, I. L. Utilization of aromatic hydrocarbons by Arthrobacter spp. Can. J. Microbiol., 13:205-211, 1967.

Stewart, F. M. and B. R. Levin. The population biology of bacterial plasmids: A priori conditions for the existence of conjugationally transmitted factors. Genetics, 87:209-228, 1977.

Stotzky, G. Activity, Ecology and Population Dynamics of Microorganisms in Soil. In: CRC Critical Reviews in Microbiology, vol. 2, A. L. Laskin and H. Lechevalier, eds., CRC Press, Cleveland, OH, pp. 59-137, 1972.

Straus, O. H. and A. Goldstein. Zone behavior of enzymes, illustrated by the effect of dissociation constant and dilution of the system cholinesterasephysostigmine. J. Gen. Physiol., 26:559-585, 1943.

Suarez, J. E. and K. F. Chater. Polyethylene glycol-assisted transfection of Streptomyces protoplasts. J. Bacteriol., 142:8-14, 1980.

Suflita, J. M., A. Horowitz, D. R. Shelton, J. M. Tiedje. Dehalogenation: A novel pathway for the anaerobic biodegradation of haloaromatic compounds. Science 218:1115-1117, 1982.

Suidan, M., W. H. Cross, K. A. Kahn and M. Fong. Treatment of phenol and substituted phenols with an anaerobic activated carbon filter. In: Chemistry in Water Reuse, W. J. Cooper, ed., Ann Arbor Science Press, Ann Arbor, pp. 509-520, 1981.

Suidan, M. T., W. H. Cross and M. Fong. Continuous bioregeneration of granular activated carbon during the anaerobic degradation of catechol. Prog. Wat. Tech., 12:203-214, 1980.

Summers, A. O. and S. Silver. Microbial transformations of metals. Ann. Rev. Microbiol., 32:637-672, 1978.

Sundar Raj, C. V., T. Ramakrishnan. Transduction in Mycobacterium smegmatis. Nature London, 228:280-281, 1970.

Sundstrom G. 3,3',4,4'-Tetrachloroazobenzene and 3,3',4,4'-Tetrachloroazoxybenzene, Potent Carcinogens and Enzyme Inducers - An Overview. In: Chlorinated Dioxins and Related Compounds: Impact on the Environment, O. Hutzinger, R. W. Frei, E. Merian and F. Pocchiari, eds., Pergamon Press, New York, pp. 337-353, 1982.

Tadros, T. F. Particle-surface Adhesion. In: Microbial Adhesion to Surfaces, R.C.W. Berkeley, J. M. Lynch, J. Melling, P. R. Rutter and B. Vincent, eds., Ellis Horwood, Ltd., Chichester U.K. pp. 93-116, 1980.

Tanaka, T. and T. Koshikawa. Isolation and characterization of four types of plasmids from Bacillus subtilis (natto). J. Bacteriol., 131:699-701, 1977.

Tempest, D. W. and O. M. Neijssel. Eco-physiological Aspects of Microbial Growth in Aerobic Nutrient-limited Environments. In: Advances in Microbial Ecology Vol. 2., M. Alexander, ed., Academic Press, New York, pp. 105-147, 1978.

Tiedje, J. M., J. M. Duxbury, M. Alexander and J. E. Dawson. 2,4D Metabolism: Pathway of degradation of chlorocatechols by Arthobacter sp. J. Agr. Food Chem., 17:1021-1026, 1969.

Timmis, K. N. Gene Manipulation In Vitro. In: Genetics as a Tool in Microbiology, Society for General Microbiology Symposium 31, S. W. Glover and D. A. Hopwood, eds., Cambridge University Press, Cambridge, U.K., pp. 50-109, 1981.

Tinsley, I. J. Chemical Concepts in Pollutant Behavior, John Wiley and Sons, New York, pp. 171-206, 1979.

Tison, D. L. and D. H. Pope. Effect of temperature on mineralization by heterotrophic bacteria. Appl. Environ. Microbiol., 39:584-587, 1980.

Tokunaga, T., Y. Mizuguchi and K. Suga. Genetic recombination in Mycobacteria. J. Bacteriol., 113:1104-1111, 1973.

Tokunaga, T. and R. M. Nakamura. Infection of competent Mycobacterium smegmatis with deoxyribonucleic acid extracted from bacteriophage Bl. J. Virol., 2:110-117, 1968.

Towner, K. J. Chromosome mapping in Acinetobacter calocoaceticus. J. Gen. Microbiol., 104:175-180, 1978.

Twarog, R. and L. E. Blouse. Isolation and characterization of transducing bacteriophage BPl for Bacterium anitratum (Achromobacter sp.). J. Virol., 2:716-722, 1968.

Ullrich, A. J. Shine, J. Chirgwin, R. Pictet, E. Tischler, W. J. Rutter and H. M. Goodman. Rat insulin genes: Construction of plasmids containing the coding sequences. Science, 196:1313-1319, 1977.

US Environmental Protection Agency. Discussion of Biodegradation in Microcosms. In: Microbial Degradation of Pollutants in Marine Environments, A. W. Bourquin and P. H. Pritchard, eds., US EPA report No. 600/9-79-012, pp. 332-336, 1979.

Van der Linden, A. C. and G. J. E. Thijsse. The mechanisms of microbial oxidation of petroleum hydrocarbons. Adv. Enzymol., 27:469-546, 1965.

Vela, G. R. and J. R. Ralston. The effect of temperature on phenol degradation in wastewater. Can. J. Microbiol., 24:1366-1370, 1978.

Veldkamp, H. and H. W. Jannasch. Mixed culture studies with the chemostat. J. Appl. Chem. Biotechnol., 22:105-123, 1972.

Vosberg, H.-P. Molecular cloning of DNA. Human Genetics, 40:1-72, 1977.

Walker, J. D. and R. R. Colwell. Mercury-resistant bacteria and petroleum degradation. Appl. Microbiol., 27:285-287, 1974.

Walker, J. D. and J. J. Cooney. Oxidation of n-alkanes by Cladosporium resinae. Can. J. Microbiol., 19:1325-1330, 1973.

Walker, N. and D. Harris. Metabolism of 3-chlorobenzoic acid by Azotobacter species. Soil Biol. Biochem., 2:27-32, 1970.

Wallace, H. R. Dispersal in Time and Space: Soil Pathogens. In Plant Disease: An Advanced Treatise. Vol. II, J. G. Horsfall and E. B. Cowling, eds., Academic Press, New York, pp. 181-202, 1978.

Wartell, R. M. and W. S. Reznikoff. Cloning DNA restriction endonuclease fragments with protruding single-stranded ends. Gene, 9:307-319, 1980.

Wedemeyer, G. Dechlorination of DDT by Aerobacter aerogenes. Science, 152:647, 1966.

Wedemeyer, G. Dechlorination of 1,1,1-trichloro-2,2-bis-(p-chlorophenyl)-ethane by Aerobacter aerogenes. Appl. Microbiol., 15:569-574, 1967.

Weiss, R. F. The temporal and spatial distribution of tropospheric nitrous oxide. J. Geophys. Res., 86:7185-7196, 1981.

Weiss, R. F. and H. Craig. Production of atmosphere nitrous oxide by combustion. Geophys. Res. Letters, 3:751-753, 1976.

Wheatcroft, R. and P. A. Williams. Rapid methods for the study of both stable and unstable plasmids in Pseudomonas. J. Gen. Microbiol. 124:433-437, 1981.

White, A., P. Handler and E. Smith. Principles of Biochemistry, 3rd ed., McGraw-Hill, New York, pp. 220-233, 1964.

Wierich, P. and P. Gerike. The fate of soluble, recalcitrant and adsorbing compounds in activated sludge plants. Ecotoxicol. Environ. Safety, 5:161-170, 1981.

Wilkinson, T. G., H. H. Topiwala and G. Hamer. Interactions in a mixed bacterial population growing on methane in continuous culture. Biotech. Bioengr., 16:41-59, 1974.

Williams, P. A. and K. Murray. Metabolism of benzoate and the methylbenzoates by Pseudomonas putida (arvilla) mt-2: Evidence for the existence of a TOL plasmid. J. Bacteriol., 120:416-423, 1974.

Wilson, J. Removal of organics from water by granular activated carbon and microorganisms. Process Biochem., 16(4):9-12, 1981.

Winogradskii, S. Bacteriologie du Sol Masson, Paris, 1949.

Wolfe, R. S. Microbial Biochemistry of Methane - A Study in Contrasts. In: Microbial Biochemistry Vol 21. J. R. Quayle, ed., University Park Press, Baltimore, pp. 268-300, 1979.

Wouters, J.T.M., F. L. Driehuis, P. J. Polaczek, M.-L. H. A. van Oppenraay and J. G. van Andel. Persistence of pBR322 plasmid in E. coli K12 grown in chemostat cultures. Antonie van Leeuwenhoek, 46:353-362, 1980.

Wright, F. C., J. C. Riner, J. S. Palmer and J. C. Schlinke. Metabolic and residue studies with 2-(2,4,5-trichlorophenoxy)-ethyl-2,2-dichloropropionate (erbon) herbicide in sheep. J. Agr. Food Chem., 18:845-847, 1970.

Yano, K. and T. Nishi. pKJ1, a naturally occurring plasmid coding for toluene degradation and resistance to streptomycin and sulfonamides. J. Bacteriol., 143:552-560, 1980.

Yayanos, A. A., A. S. Dietz and R. van Boxtel. Obligately barophilic bacterium from the Mariana Trench. Proc. Nat'l. Acad. Sci. U.S., 78:5212-5215, 1981.

Yeats, S., P. McWilliam and W. Zillig, A plasmid in the archaebacterium Sulfolobus acidocaldarius. EMBO J., 1:1035-1038, 1982.

Yeoh, H. T., H. R. Bungay and N. R. Krieg. A microbial interaction involving combined mutualism and inhibition. Can. J. Microbiol., 14:491-492, 1968.

ZoBell, C. E. The effect of solid surfaces upon bacterial activity. J. Bacteriol., 46:39-56, 1943.

Zuniga, M. C., D. R. Durham and R. A. Welch. Plasmid and chromosome-mediated dissimilation of naphthalene and salicylate in Pseudomonas putida PMD-1. J. Bacteriol., 147:836-843, 1981.

Glossary

activated sludge - the concentrated, settled microorganisms from a forced-aeration basin. In the activated sludge process, these microorganisms are pumped into wastewater entering an aeration basin to provide a high biological oxidative potential; they are collected again by sedimentation as the wastewater leaves the basin.

aeration basin - any device aiding the aeration of a wastewater to accelerate the rate of biological oxidation of pollutants.

archaebacteria - a group of microorganisms comprising one of the two primary kingdoms of prokaryotes, the other being the eubacteria. Archaebacteria have cell walls containing isoprenoid lipids. Some archaebacteria also contain unique enzyme cofactors. Typical archaebacteria include methanogenic bacteria, certain halophiles and certain extreme thermophiles/acidophiles.

autoradiography - a method for detecting a radioactive material by laying a photographic film directly onto an object in the dark, allowing for exposure of the film by the radiation, followed by development of the film to visualize the image of the radioactive source.

autotroph - an organism that obtains its nourishment (carbon and energy) only from inorganic matter (chemoautotroph), or from inorganic matter and light (photoautotroph).

bacteriocins - proteinaceous antibiotics produced by bacteria.

biochemical oxygen demand (BOD) - the weight of oxygen that can be consumed by biological oxidation of pollutants in a wastewater, per unit volume of the water, usually in a fixed period of time.

bulking - abnormal and poor clarification of a wastewater by sedimentation, usually due to the poor flocculation and settling characteristics of the sludge microorganisms present in it.

catabolism - the degradative metabolism of a compound usually to yield energy or fundamental chemical units. The opposite of anabolism.

cDNA - see complementary DNA

clone (noun) - an identical copy of a biological structure, usually obtained by allowing self-replicating entities to grow. Clones commonly refer to a genetically identical collection of bacteria all descendent from a single, original bacterium, or to a fragment of DNA ligated to a self replicating DNA fragment (cloning vector) and reproduced in a living cell.

clone (verb) - in molecular biology, to reproduce a fragment of DNA by attaching it to a segment of DNA called a cloning vector, that has a base sequence permitting its replication in a cell. When in a cell, the combination of vector and "foreign" DNA is reproduced as a unit under control of the vector.

co-metabolism - the transformation of a non-growth substrate in the obligate presence of a growth substrate or another transformable compound. A non-growth substrate is one that cannot support cellular division, although it may support an increase in cell mass. See Dalton and Stirling (1982) for detail.

complementary DNA (cDNA) - a strand of DNA synthesized on an RNA template by the enzyme reverse transcriptase.

complementation - the interaction of the products of two genes to restore the functionality of a set of enzyme reactions in a cell. The term refers to two common experiments: Originally, it referred to a test to define whether two independent mutations giving similar phenotypes were in the gene for the same gene product. A strain was constructed containing a copy of the mutant gene from each source. If a wild type phenotype was exhibited by this strain, the mutations were said to complement each other. In this case the mutations resided in genes for different gene products, the homologous, normal genes linked to each mutant gene producing normal gene products and restoring the functionality of the pathway. If the two mutations resided in the gene for the same gene product, no functional gene product would result and the mutations were called noncomplementary. The second type of experiment identified the presence of a gene for a functional equivalent of a mutated gene on a DNA fragment from a foreign source. Usually such experiments involve the introduction of a recombinant plasmid or other cloning vector bearing foreign DNA into a bacterial host cell mutant in some enzymatic step. If the recombinant fragment restores the wild type phenotype, it must produce a functional equivalent of the mutant host enzyme and it is, therefore, said to complement the mutation. This is a very useful means to select desired DNA fragments in shotgun cloning experiments.

conjugation - the transmission of DNA between two bacteria by cell-to-cell contact. In all known cases, such transmission depends on enzymatic functions encoded on certain plasmids called conjugative plasmids. The DNA transmitted between cells can be plasmid DNA and, in some cases, plasmid and chromosomal DNA.

consortium - a two-membered culture or a natural assemblage of microorganisms in which each organisms benefits from the other. Consortia based on metabolic interactions are frequently isolated by enrichment culture.

copiotroph - see eutroph

deoxyribonucleic acid (DNA) - a polymer made up of 3' to 5' phosphodiester linkages between 5'-deoxyribomononucleotides. Certain DNA nucleotide bases are said to be complementary because they can hydrogen bond to each other to give constant spacings between two single strands of DNA. If two DNA strands have a sequence of bases whose order is mirrored by the exactly complementary bases in another strand, then these two strands can be joined by base hydrogen bonding to produce a regular double stranded structure. Each strand of a double stranded DNA can be exactly replicated by breaking the hydrogen bonds between bases and synthesizing a new DNA strand by linking together nucleotides that are exactly complementary to the exposed bases on the opened single DNA strands. The sequence of bases in DNA is able to encode information interpretable by the cellular machinery and DNA is the ultimate repository of genetic information in a cell.

diauxie - sequential consumption of the nutrients available to a microorganism; the utilization of an available nutrient only after the supply of another available nutrient has been exhausted.

DNA - see deoxyribonucleic acid.

elective culture - see enrichment culture.

endonuclease - an enzyme hydrolyzing the phosphodiester backbone of a nucleic acid anywhere but at the ends of the molecule, producing a hydroxyl- and a phosphate-ended oligonucleotide.

enrichment culture - a means of isolating microorganisms with specific catabolic potential. A mixture of microorganisms, for instance, a portion of sediment, soil, or wastewater sludge, is used to inoculate a sterile medium containing a specific nutrient. After a period allowing for growth at the expense of the nutrient, an aliquot of this culture is used to inoculate fresh medium of the same composition, and this cycle is repeated several times. The process dilutes the number of organisms unable to reproduce at the expense of the nutrient, enriching for those that can. Also called elective culture.

eutroph - an organism colonizing habitats rich in nutrients (eutrophic environments). Also called copiotrophs by Poindexter (1981a).

exconjugant - any of the cells in a culture assembled to allow the transmission of genes among bacteria by conjugation. Exconjugants can be unchanged parental bacteria or transconjugants.

exonuclease - an enzyme that hydrolyzes the phosphodiester backbone of a nucleic acid only at a terminal nucleotide, producing a mononucleotide and a nucleic acid smaller by one nucleotide.

F' - an F (fertility) plasmid carrying additional DNA that encodes gene(s) not normally present on the F plasmid.

flocculation - the interaction of microscopic solids tending to link them together and enhancing their ability to settle by gravity.

genome - the total genetic information contained in a cell.

genotype - the specific genetic structure underlying a particular phenotype. Usually this refers to a specific set of mutations present in a strain. Many different genotypes in different stains may give rise to the same phenotype.

heteroduplex - the structure formed by hybridization of two specific DNAs sharing significant regions of homology. Heteroduplexes are usually prepared under conditions enabling the subsequent examination of individual heteroduplexes in the electron microscope, to visualize and to map the location of the homologous, double-stranded regions and the heterologous, single-stranded regions on each of the starting DNAs.

heterotroph - an organism that obtains its nourishment (carbon and energy) from organic matter.

Hfr - high frequency recombination. A bacterial strain in which a fertility plasmid has inserted into the chromosome thereby enabling the strain to transfer its chromosomal genes to recipient strains at high frequency.

hybridization - the association of complementary base sequences on single-stranded DNAs from different sources, to form a region of double-stranded DNA. Coupled with a means to detect these double-stranded regions, hybridization offers a means to test different DNAs for homology.

incompatibility - the failure of two plasmids to be stably maintained in a bacterium. During the division of a bacterium originally harboring two incompatible plasmids, the plasmids will rapidly segregate so that most daughter cells will harbor one or the other but not both of the plasmids. Incompatibility is thought to reflect a competition between the plasmids for some commonly needed factor essential to plasmid maintainance in the cell. Hence, incompatible plasmids are considered to be similar and incompatibility is used to classify plasmids into groups. Incompatibility groups contain mutually-incompatible plasmids all designated by the prefix "inc" followed by a letter or a letter and number, e.g., inc B and inc P-2. Plasmids in different incompatibility groups are stably maintained in a bacterium and do not rapidly segregate during cell division.

insertion sequence (abbrev. IS) - certain DNA base sequences bounded at their ends by repeated sequences and able to insert into replicating DNA, interrupting the previous DNA base sequence at the point of insertion. Insertion sequences have been found in plasmids, in phage genomes, in bacterial chromosomes and at the termini of some transposons. Some of the first recognized insertion sequences have been extensively characterized. They range from 0.7 to 5.8 kb in size and have been given number designations, e.g., IS1, IS2, etc.

insertional inactivation - the interruption of a gene's base sequence by a transposable element leading to the loss of production of a functional form of the gene product.

in situ - lit. in place. Occurring at the original site or location.

intercalation - insertion, interposing of one entity into another. In molecular biology, the insertion of planar, aromatic compounds, between stacked bases in DNA.

in vitro - lit., in glass. Occurring under artifically constructed experimental conditions, for instance in a test tube.

in vivo - in a living organism.

IS - see insertion sequence.

leachate - liquid material escaping from a landfill or other repository of solid waste, often produced by the percolation of rainwater through the waste.

ligase - in molecular biology, usually refers to DNA ligase, an enzyme catalyzing synthesis of an ester linkage between the 3'-hydroxyl and 5'-phosphate ends of DNA. Usually such ends are in close proximity due to hydrogen bonding of the rest of the substrate DNAs to a complementary DNA strand that bridges the site of esterification. Unlike other ligases, T-4 DNA ligase can simultaneously join the appropriate ends of blunt-ended double stranded DNA. In general, ligases are a class of enzymes catalyzing the linkage of two substrate molecules, not necessarily just DNAs.

ligation - in molecular biology, the joining of two DNA strands, usually by a DNA ligase enzyme. In general, the linking together of two substrate molecules.

multinucleate - having more than one nucleus in a cell. A nucleus is the collection of chromosomes bounded by a nuclear membrane and usually distributed one to a cell. Multinucleate cells arise by the fusion of protoplasts, by syncytial grown habit, or in other ways.

mutagen - a chemical or physical agent able to induce mutations in an organism.

mutagenesis - the creation of mutation(s) in an organism by treatment with chemical or physical agents called mutagens. Usually such agents damage DNA but they may have other modes of action. After treatment, a period of growth is needed to allow misrepair of the damaged DNA, leading to an alteration in the information encoded in the DNA.

mutation - a change in the information encoded in the genetic apparatus of a cell. In microorganisms mutation is usually the result of an alteration of the base sequence in DNA.

oligotroph - an organism colonizing habitats sparse in nutrients (oligotrophic environments).

operator - that part of an operon to which the repressor protein binds. When bound to the operator, the repressor blocks the transcription of the operon's structural genes.

operon - an assemblage of structural genes and closely linked special regions of DNA that enable the regulation of gene transcription. One special region, termed the promoter, is the site of binding of DNA-dependent RNA polyerase. Between the promoter and the structural genes is a special region called the operator, the binding site of a regulatory protein called the repressor. When bound to the operator, the repressor prevents RNA polymerase from transcribing the structural genes. Opposite the promoter-operator side of the structural gene(s) is a region of DNA that causes the termination of transcription. Some operons also contain additional special regions affecting the control of transcription by other means.

palindrome - any sequence of informational units reading the same back-to-front and front-to-back, for instance: MADAM I'M ADAM. In molecular biology, a palindrome is a DNA base sequence showing a two-fold symmetry around a point, such as:

5' - GAG CTC - 3'
3' - CTC·GAG - 5'

phenotype - the observed characteristics exhibited by a living organism, for instance, a bacterium able to grow at the expense of glucose and inorganic nutrients only in the presence of the amino acid histidine, due to a defect in histidine biosynthesis.

plaque - a small, clear region in a lawn culture of bacteria, caused by the lysis of the bacteria. Lysis is usually the result of infection of the bacteria by bacteriophage.

polar mutation - a mutation that reduces the transcription of unmutagenized genes linked to the mutated gene and transcribed after it. Mutations caused by insertion of a transposon into a gene are frequently polar.

polychlorinated biphenyls (PCB) - a collection of homologous compounds produced by the chlorination of biphenyl. PCB's have wide industrial uses because they are electrical insulators yet they have good thermal conductivities and are relatively chemically inert.

primary treatment - the clarification of influent wastewater by sedimentation.

promoter - a region of DNA to which DNA-dependent RNA polymerase binds.

protoplast - a cell bounded by a cytoplasmic membrane, but lacking any rigid layer. Bacterial protoplasts usually are formed by lysing the peptidoglycan layer of Gram positive bacteria in an osmotically protective medium, and allowing escape of the cell membrane-enclosed body from the rigid cell wall remanent.

recombinant DNA - see recombination.

recombination - the reassortment of genetic traits due to a breakage and rejoining of DNA. The product of recombination can be either a recombinant organism, or a recombinant DNA.

replicase - an enzyme catalyzing the duplication of a DNA by synthesis of a new DNA strand from nucleotide monomers using the strand to be duplicated as a template.

replicon - any autonomously replicating unit of DNA. Bacterial cells often contain more than one replicon, for instance a chromosome and a plasmid.

restriction - in molecular biology, the hydrolysis of foreign DNA by certain endonucleases that recognize specific short DNA base sequences. Originally, the term referred to the failure of certain bacteriophage to give productive infection depending upon the host from which they came. This was later found to be due to the action of restriction endonucleases. Restriction endonucleases fail to hydrolyze DNA in which certain bases in the sequence that they recognize are methylated (modified). Thus, if the original host of a bacteriophage methylated the bacteriophage DNA at the sites recognized by the new hosts' restriction enzymes, the phage DNA would not be attacked and a productive infection could occur. Appropriate methylation is thus a characteristic differentiating DNA's into self-like or non-self-like groups, and restriction/modification enzymes are a means for protecting a cell from foreign, non-self DNA.

reverse transcriptase - an enzyme that synthesizes a strand of DNA complementary to the base sequence of an RNA template.

rhizosphere - the region of soil penetrated and influenced by plant roots.

ribonucleic acid (RNA) - a polymer made up of 3' to 5' phosphodiester linkages of 5'-ribomononucleotides. RNA often functions in living cells as intermediate carriers of information encoded in DNA, or as part of the cellular machinery that translates genetic information into the functional, proteinaceous structures that mediate cellular activities.

RNA - see ribonucleic acid

secondary treatment - the aerobic biological digestion of organic pollutants in a wastewater, usually performed in a reactor that assists aeration (aeration basin). Such reactors include trickling filters, activated sludge reactors and rotating biological disks.

sedimentation basin - a device slowing the rate of flow of a wastewater to a point where suspended, entrained matter settles out. The device usually also permits the settled material to be continuously collected for disposal.

site-specific mutagenesis - a treatment with a chemical or physical agent that leads to changes in the information encoded in only very small specific regions of DNA. This is usually accomplished using agents that act only on particular nucleic acid architectures, such as the single-stranded portion of a "D" loop, or at a nick in double stranded DNA. These forms of DNA in turn are created at specific sites using sequence-specific probes or enzymes.

sludge - any of various mud-like deposits that settle from slow-flowing wastewater. The composition of a sludge will depend upon the source of the wastewater. Sludges can be made up primarily of inorganic precipitates, insoluble organic matter or microorganisms. See activated sludge.

sludge age - in wastewater treatment, the ratio of the total biomass in a reactor to the rate of production (or loss) of biomass in the reactor. (Under steady state operation, the rate of biomass production equals the rate of biomass loss). Sludge age is approximately the reciprocal of the net growth rate.

spheroplast - the spherical body of a bacterium that is produced by hydrolyzing the rigid peptidoglycan layer of the cell wall, usually of Gram negative bacteria. Spheroplasts retain much of their outer cell wall layers, including some lipopolysaccharide and the broken remenants of the peptidoglycan. See protoplast.

syncyticum, n. (adj., syncytial) - that structure in filamentous microorganisms that is produced when incomplete septa form between cells, causing the protoplasm to be continuous, and the structure to be multinucleate. Also, any multinucleate mass of protoplasm.

T4 - a bacteriophage infecting *E. coli*.

TCDD - see tetrachlorodibenzodioxin

tertiary treatment - any wastewater treatment in addition to sedimentation (primary treatment) and aerobic, biological (secondary) treatments. Tertiary treatments may include anaerobic biological treatment, physical treatments such as adsorption, or chemical treatments.

2,3,7,8-tetrachlorodibenzodioxin (TCDD) - an extremely persistent and toxic compound formed as a by-product of the manufacture of the herbicide, 2,4,5-trichlorophenoxyacetic acid and the combustion of a variety of organochlorine compounds. TCDD also signifies a number of highly toxic chlorinated dibenzodioxins and dibenzofurans co-occurring with TCDD.

transconjugant - a bacterium that has received new DNA from a DNA donor by conjugation. Transconjugants may be genetically unstable, that is, the genes on the newly acquired DNA may not be uniformly inherited by the daughters of the transconjugant. Genetically stable descendants possessing some or all of the genes acquired by a transconjugant are called recombinant strains.

transcription - the synthesis of RNA from a DNA template by DNA-dependent RNA polymerase.

transduction - the transmission of non-viral DNA among bacteria via virus particles. Transduction can result when host cell DNA rather than viral DNA is incorporated into reproducing virus particles. Subsequent infection of a cell with such particles leads to injection of the host cell DNA into a new bacterial cell where it is incorporated into the genome by normal recombination.

transfection - the uptake of isolated viral DNA by a cell resulting in the reproduction of the virus. The transformation of a cell using viral DNA.

transformation - the uptake and incorporation of isolated DNA by a cell.

translation - the synthesis of a polypeptide from an RNA template by a ribosome.

transposable element - any specialized DNA base sequence capable of insertion into a large number of sites in replicating DNA. Recognized transposable elements include transposons, insertion sequences and mu phage.

transposition - the insertion of a specialized DNA base sequence called an insertion element or a transposon into a replicating DNA.

transposon - a specialized DNA base sequence (1) that is able to be inserted by an enzyme called transposase into a large number of specified sites in a replicating DNA and (2) encoding the gene for a selectable phenotype, for example, a gene for an antibiotic resistance determinant.

transposon mutagenesis - the production of a mutation by the transpositional insertion of a DNA segment, a transposon, into a gene. Such insertion divides the gene's original base sequence into two separate parts.

Index

Other Noyes Publications

APPLIED GENETIC ENGINEERING
Future Trends and Problems

by
Morris A. Levin
U.S. Environmental Protection Agency

Robert H. Zaugg
Sandoz, Inc.

Jeffrey R. Swarz
Pall Corporation

George H. Kidd
International Plant Research Institute

This concise, up-to-date study details industrial and agricultural applications of genetic engineering and identifies future trends and problems which may result. Few areas today evoke greater interest in the scientific and lay communities, or hold greater potential, than the field of applied genetics. The book elucidates many of the existing possibilities as well as some of the major stumbling blocks which face scientific personnel and corporate and academic decision makers as we enter this age of molecular biology and genetic manipulation.

Following a brief history of events leading to the current boom in genetic research, are descriptions of genetic engineering applications in key industrial sectors. These are supplemented by discussions on directions for research, potential products, environmental and health concerns, and the possibilities for joint efforts between academic institutions and/or corporate enterprises.

Key research personnel and their academic or corporate affiliations are identified. Realistic appraisals of various technologies, including time perspectives, are provided. A discussion of past, present, and anticipated regulatory activity is also included. Finally, the book contains extensive tables listing biotechnology companies, corporations and universities with major commitments to genetic research, federal government activities, international projects, and a compilation of current newsletters in the field.

A condensed table of contents including **chapter titles and selected subtitles** is given below.

ISBN 0-8155-0925-1 (1983) **191 pages**

Other Noyes Publications

GENETIC ENGINEERING APPLICATIONS FOR INDUSTRY

Edited by J.K. Paul

Chemical Technology Review No. 197

The documents that make up this book are four working papers which were prepared by *Genex Corporation, Poly-Planning Services,* and the *Massachusetts Institute of Technology* for the U.S. Congress's Office of Technology Assessment (OTA) report on the impacts of genetic engineering. **These papers contain significantly more technical information than is found in the OTA report.** They are published in their entirety.

Genetic engineering, with all of its implications and ramifications, is the most exciting area of scientific research in the world today.

Applications of genetic engineering offer almost unlimited challenges in such industries as chemicals, foods, pharmaceuticals, agricultural products, and energy, and, of course, in medical research. The book includes technical discussions of previous accomplishments and potential research directions for the future.

A **condensed table of contents, with selected subtitles,** is given below.

ISBN 0-8155-0869-7 **580 pages**